Lyrical
Life Science

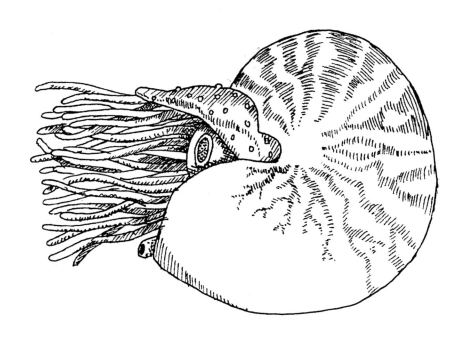

**Text and Lyrics by
Doug C. Eldon**

**Performed by
Bobby Horton**

**Illustrations by
Eric Altendorf**

Published by:
Lyrical Learning
8008 Cardwell Hill Dr.
Corvallis, OR 97330
541-754-3579

> These book and CD sets can be used by students of many ages. Younger elementary students can benefit by becoming familiar with scientific terms through casual listening, yet may not fully understand the information until later. These songs, texts, and workbooks were written with middle school students in mind, as an introduction to life science. The information presented in these three volumes of life science should be very familiar to high school students before they study biology in greater depth.
>
> Workbooks are also available for each volume. Ideally, these resources should be in addition to hands-on activities where observations and applications can be made. In this way the knowledge learned through singing and reading can also become known through experience.
>
> Our website www.lyricallearning.com has reviews and awards; FAQs; an article about the theory and research behind using lyrics and music for learning; information about our products as well as Volume 1 revisions and updates; a list of our distributors and ordering information; and where to find our products as digital downloads.

Text and lyrics: Doug Eldon
Layout and design: Dorry Eldon
Cover art and design: Susan Moore
Illustrations: Eric Altendorf
Sheet music layout: Mike Martin and Art Baines
Scientific advisors: Steve Lebsack and Dr. Karen Timm
Special thanks to: Dr. Brad Smith, Bo Blair, MaryAnn Brown, Sara Cooper, Pat Eide, Anne Hetherington, Cathy Passmore, and Roberta Sobotka.

Illustrations redrawn with permission:

Pimentel, Richard A. *Invertebrate Identification Manual*, American Institute of Biological Sciences, 1967. Figures on pages 19 – 25.

Swanson/Webster, *The Cell*, Prentice Hall. Upper Saddle River, New Jersey, 1995. Figure on page 55.

Weier, T. Elliot, et al, *Botany: An Introduction, 4th Edition*, John Wiley & Sons, Inc. Figures of apple blossom, maturing apple, mature fruit, and grass flower on pages 44 – 45.

Copyright Lyrical Learning
   1st printing 1995
   Reprinted annually through 2010

Volume 1 Text, CD and workbook set: ISBN 0-9741635-4-6

TABLE OF CONTENTS

List of Scientific Illustrations		2
Introduction		3
Chapter 1	**The Scientific Method**	4
	Six Steps of the Scientific Method,	
	The Scientific Method in Use,	
	Louis Pasteur	
Chapter 2	**All Living Things**	10
	Characteristics of Living Things,	
	Needs of Livings Things,	
	The Cell Theory, Classification,	
	Levels of Organization	
Chapter 3	**Invertebrates**	18
	Common Phyla, Metamorphosis,	
	Characteristics of Common Insect Orders	
Chapter 4	**Coldblooded Vertebrates**	26
	Fish, Amphibians, Reptiles	
Chapter 5	**Birds**	30
	Characteristics, Behavior,	
	Classification	
Plants: An Introduction		34
	Vascular Tissue, Reproduction,	
	Classification	
Chapter 6	**Algae, Fungi and Nonvascular Plants**	36
	Kinds of Algae, Fungi, Symbiosis,	
	Lichen, Moss, Liverworts	
Chapter 7	**Vascular Plants**	42
	Vascular Tissue, Classification,	
	Gymnosperms, Angiosperms,	
	Parts of Plants, Reproduction,	
	Photosynthesis, Respiration,	
	Plant Growth,	
	Common Angiosperms	
Chapter 8	**Protozoa**	48
	Sarcodines, Ciliates,	
	Flagellates, Sporozoans	
Chapter 9	**Genetics**	52
	Traits, Genes, DNA,	
	Genetic Engineering	
Chapter 10	**Viruses**	56
	Infectious Diseases,	
	Characteristics of Viruses,	
	Viral Diseases, Reproduction	
Chapter 11	**Bacteria**	62
	Reproduction, Classification,	
	Importance, Helpful Bacteria,	
	Bacteriology, A Short History,	
	Bacterial Diseases	
Appendix		
	Lyric Sheets	75
	Classification of Animals Chart	86
	Career Search	87
	Notes	88
	Bibliography	89
	Index	90

LIST OF SCIENTIFIC ILLUSTRATIONS

Apple	44, 45
Aerie	32
Air sac system	31
Algae	34, 35, 37
Alligator	29
Amoeba	49
Bacilli	64
Bacteria	63
Bacteriophage	58
Bean plant	46
Binary fission	63
Birds	31 - 33
Bluegill	28
Cell, animal	13
Cell, plant	14
Cells, kingdom classification	14
Centipede	23
Chromosomes	54
Cocci	64
Cockroach	25
Cone	4
Crab, rock	22
Crane fly	15
Crayfish	22
Crocodile	29
Daphnia	22
Decomposer	12
Diatoms	11, 37
Disc flower	47
DNA molecule	55
Double helix	55
Douglas fir	44
Dragonfly lava	24
Duck, Mallard	33
Earthworm	20
Egg and sperms	11
Euglena	50
Feathers, contour and down	31
Fern	43
Flower	44, 45, 47
Food chain	12
Frog	28
Fungi	38, 39, 41
Grass flower	45
Grasshopper, short-horned	25
Heron	33
Housefly	16
Hydra	11
Kelp	34, 35, 37
Lacewing	12
Lamprey	27
Leaves	44, 45, 46
Lichen	40, 41
Liverworts	40, 41
Lizard	29
Metamorphosis	24
Microscope, Leeuwenhoek's	66
Millipede	23
Mold, bread	39
Mosquito	51
Moss	40, 41
Moth	24
Mushroom	12, 38, 39, 41
Nautilus, chambered	21
Oak	46
Octopus	21
Paramecium	50
Passenger Pigeon	33
Pasteur	9
Pasteur's swan-neck flask	8
Peas	53
Penguin	31
Pill bug	22
Planaria	20
Poplar	46
Portuguese man-o'-war	20
Producer	12
Protozoa	49 - 50
Queen Anne's Lace	47
Ray flower	47
Robin	11
Root systems	45
Rot	38, 71
Rush	11, 41
Salamander life cycle	28
Scientific tools	7
Scorpion	23
Sea anemone	19
Shark	27
Shell, tower	21
Silverfish	24
Slug	21
Snake	29
Sow bug	22
Spider, crab	23
Spirilla	64
Squid, giant	21
Staghorn coral	19
Stentor	50
Stingray	27
Sunflower	47
Toad	28
Tortoise	29
Tree sponge	19
Turtle	29
Vascular tissue	43
Vertebrae	27
Viruses	58, 60, 61
Volvox	11, 50
Water strider	25

INTRODUCTION

The concept of using lyrics and music to teach and to learn is as ancient as education itself. Songs, ballads, chants and poems have long been helpful in communicating ideas and information. We remember the alphabet song and with little effort remember tunes heard on the playground and around the campfire.

I began using songs to help my sixth grade students with our study of life science. Several students were having difficulty with the amount of new information. Even with visuals, hands-on activities and field trips, I could sense that part of their problem was learning the structures, functions and characteristics described in the language of science. They needed to have the foundation of factual information established in their minds before they could confidently proceed in their studies to classify, compare and analyze life science.

I condensed the most important information from a chapter, section or unit of a textbook and set it to singable familiar melodies. I discovered that by putting the scientific language to music the students found learning to be easier and more enjoyable. Other teachers in the field have had the same experience with scientific information set to music; Tom Evans, now Professor of Education at Oregon State University remarked: "The singing of songs in my middle school science classes was an immediate success. Students became more interested and actively involved in learning science."

In researching the concept of what I call lyrical learning, I found that not only has it been used for a very long time, but that a better understanding of how the brain processes, stores and retrieves information has now made it clear why this way of learning is effective:

1 - singing reinforces what is heard or read, actively involving the whole brain—the left side that deals with language and the right side that deals with music and artistic perception;
2 - the association of new information with familiar melodies makes learning easier;
3 - repetition and rehearsal become recreational, and the redundancy of practicing enjoyable;
4 - rhythms and rhymes make a memorable pattern that is easily recalled;
5 - the information grouped as lines or verses of lyrics becomes visually more memorable than sentences in paragraphs;
6 - the novelty of songs increases emotional involvement and heightens attention.

The lyrics of each tune are content-rich and loaded with scientific terms and concepts that are explained more fully in the text. Illustrations, historical notes and fun facts such as world records are added for further clarification and interest. On the accompanying cassette Bobby Horton takes these educational tunes to a new level of enjoyment and entertainment. All this is for students to learn with confidence to enjoy their scientific world in a new and refreshing way. I hope you enjoy singing these tunes and find life science as fascinating as I have.

Doug Eldon

THE SCIENTIFIC METHOD
(to the tune of "Dixie")

Dan Decatur Emmett
Lyrics by Doug Eldon

It may not seem important to you but the first thing that they always do
Is state the problem or ask a question so they know just what they're after
Then they review everything involved that might help get the problem solved
By reading, researching and gathering information

After both of these steps they take they go ahead and then they make
An educated guess; a hypothesis; a possible solution
Then they use scientific tools to measure and test some variables
In experiments which are really meant to give more information

This information they call data, they put together so that later
They can analyze and synthesize, to see just what it all means
Only when they have done all these experiments testing hypotheses
Which may prove or else disprove, then they'll state their conclusion.

This is the systematic way a scientist may use any old day
'Cause it's methodical and it's logical the scientific method.

THE SCIENTIFIC METHOD

Dr. Karen Timm is a scientist by profession who is in this book to help with the scientific information. Specifically, she is a veterinary professor and a specialist in laboratory animal medicine. She enjoys her subject so much that she raises Macaws, a species of beautiful tropical birds.

Doug Eldon is a teacher who enjoys life science almost every day. When he's not in his classroom, he raises silkworms, grows fungi and explores creeks and woods searching for insects and unusual plants. He joins Dr. Timm and adventures throughout the scientific information to show you, the reader, that life science can be very interesting and exciting.

The word **science** means to have knowledge of, or to know. It also means the organized system of knowledge that comes from observation, study and experimentation.

Many scientists spend their time making observations, asking questions and finding problems to solve. In fact, part of the reason they are scientists is that they are good at problem solving and enjoy the challenge of figuring things out. They study and think about what is known and what needs to be learned. Then they may experiment, and use the information gained to answer the question or solve the problem they have. This whole process has been called the **scientific method**.

If you study any particular field or branch of science, you will find that there are facts, principles (rules) and methods (ways of doing things). There are also many words or terms that help scientists communicate what they know. One aspect of understanding science is learning the vocabulary; another is learning the systems for organizing information; and still another part is learning to think and to solve problems the scientific way!

A systematic way of doing things is organized and purposeful; it is methodical if it involves orderly steps; and it is logical if it is reasonable, makes sense and is well thought out beforehand. The scientific method combines all of these aspects.

Six steps often included in the scientific method are:

#1 - Ask a question or state the problem
2 - Gather information
3 - Make a hypothesis
4 - Experiment
5 - Analyze data
6 - State a conclusion

THE SCIENTIFIC METHOD IN USE

To demonstrate how the scientific method can be used, let's go back to the 1860's when a French scientist named Louis Pasteur was trying to solve one of the most basic questions in the study of life: Where do living things come from?

We know now that living things come from other living things but scientists in earlier times didn't have the knowledge or understanding about how things came to be. Their explanation often was that life occurred all by itself. Nonliving matter was thought to produce living things: frogs were thought to come from the mud of riverbanks, beetles from dung and maggots from decaying meat! This simple explanation became known as the theory of **spontaneous generation**.

Meat in open container was exposed to flies—maggots appeared on the meat

Meat in covered container was not exposed to flies—no maggots appeared

Step 1 - Ask a Question or State the Problem

For Pasteur the question was: Do living things come alive spontaneously, or do they all come from parents of the same kind? He was at the time working with microscopic organisms, and so his specific question was: Do germs just appear from nowhere?

Step 2 - Gather Information

Pasteur studied the works of other scientists. He read of the experiment an Italian biologist named Redi had done in 1668 that proved that maggots did not come from meat. If flies were kept off the meat by covering it, the fly eggs could not be laid and maggots did not appear. Pasteur learned from this experiment that Redi convinced scientists that complex creatures came only from others like themselves. But what about microscopic forms of life?

The microscope was first developed and used to see tiny organisms in the early 1700's but it was not until about 1760 that another Italian biologist named Spallanzani seriously began to study microorganisms. Pasteur read of his experiments which showed that microscopic organisms, or **microbes**, did not come to life spontaneously, but were often carried by dust in the air. Spallanzani also discovered that microbes reproduced by dividing, and were killed by boiling.

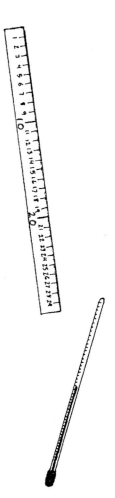

Scientists use metric tools when conducting experiments

Step 3 - Make a Hypothesis

Pasteur used what he learned from Spallanzani and Redi and what he knew from his own experiments to form a **hypothesis** (a possible answer to his question). He basically restated much of what Spallanzani had said one hundred years earlier. That was:

    1 - living things come from other living things of the same kind.
    2 - some are too small to be seen without a microscope.
    3 - they are spread by floating on dust in the air.

Step 4 - Experiment

Spallanzani conducted an experiment which involved boiling a liquid in a glass flask, then sealing it. He found that microbes did not appear in the closed flask, but grew in an open one. He concluded from this experiment that microbes are in the air and boiling kills them. Other scientists still argued that the boiling changed the conditions needed for spontaneous generation to occur and closing the flasks kept oxygen out, which the microbes needed to live and grow. Factors that might affect the results of an experiment are called **variables**. Pasteur had to design an experiment where he could control all the variables in order to prove that microbes indeed were carried on dust in the air, and did not just appear.

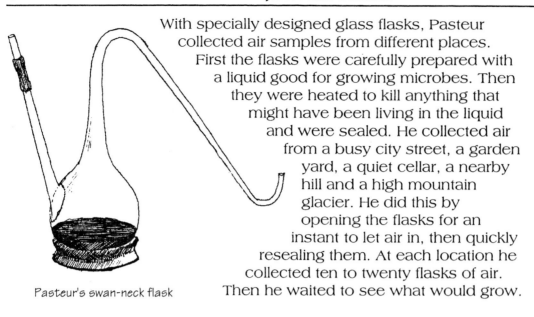

Pasteur's swan-neck flask

With specially designed glass flasks, Pasteur collected air samples from different places. First the flasks were carefully prepared with a liquid good for growing microbes. Then they were heated to kill anything that might have been living in the liquid and were sealed. He collected air from a busy city street, a garden yard, a quiet cellar, a nearby hill and a high mountain glacier. He did this by opening the flasks for an instant to let air in, then quickly resealing them. At each location he collected ten to twenty flasks of air. Then he waited to see what would grow.

Step 5 - Analyze Data

Pasteur could have recorded his data in a table such as this:

Location of air sample	Number of flasks	Number with microbes
city street	10	10
garden yard	11	11
quiet cellar	10	1
nearby hill	20	8
high mountain	20	1

By **synthesizing** or putting together his data, Pasteur could clearly see his results. Pasteur reasoned that the air from the city and the yard both had a lot of microbe-carrying dust, but that the air from the cellar was quite still, with little dust. He further reasoned that the air in higher elevations had less and less dust and therefore fewer and fewer microbes.

Step 6 - State a Conclusion

Pasteur learned from experiments others had done; he made a hypothesis; he then performed his own experiments, controlling the variables; and he analyzed the data he gathered. He could now conclude that the dust in the air carried microbes and they did not just appear but come from other microbes of the same kind.

Although the scientific method was not something formally taught or learned by Pasteur, it was the process he used. Scientists continue to use this approach to solve modern scientific problems.

Many scientists still argued for spontaneous generation. After many years of repeating his experiments, Pasteur finally convinced the scientific world that he was right! As you will see in the chapters on viruses, bacteria and protozoa, what became known as the **germ theory** changed science and medicine in many ways.

LOUIS PASTEUR
1822-1895

ALL LIVING THINGS
(to the tune of "I Love the Mountains")

Traditional
Lyrics by Doug Eldon

All living things are able to reproduce.
Move and grow and respond to stimulus...
and carry on metabolic activities ---
These are characteristics of living things.

There are four metabolic activities:
Ingestion and digestion are two of these
Respiration and excretion
Metabolic, metabolic: chemical activities

Needs of living things include energy
Water, oxygen and food to eat
Living space and proper temperatures
All living things have these six basic needs

Living things are all made up of cells
Units of structure and functions you can tell
All cells come only from other living cells
This is what's called the cell theory

Cells that are similar joined together form tissues
Tissues working together form organs
Organ systems and organisms are:
Five levels in which living things are organized

Kingdom, phylum, class, order, family
Genus and species make the name you see
Nomenclature and taxonomy
Classify, classify, name and classify

Diatom

ALL LIVING THINGS

The study of life science involves learning what it means to be alive. This may seem obvious, but for example, when viruses were discovered it was found that they were different from bacteria—they could not be killed by antibiotics because they were not really alive! Scientists have determined five characteristics of life and several processes involved in staying alive. A living thing must be able to:

1 - reproduce itself, either **sexually** (involving a female egg and male sperm) or **asexually** (involving only one parent, such as by budding).

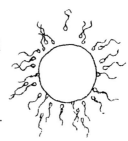
Sexual reproduction: Egg with many sperm

Western white pine

2 - move in order to meet a need such as finding shelter or protection, or hunting for food.

3 - grow and develop either in size or complexity, or both.

4 - respond to a stimulus, which means that the living thing acts in a particular way because of a change in its surroundings.

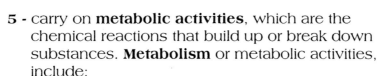
Asexual reproduction: Fresh water Hydra reproduce by budding

5 - carry on **metabolic activities**, which are the chemical reactions that build up or break down substances. **Metabolism** or metabolic activities, include:

    1 - ingestion: the eating or absorbing of food.
    2 - digestion: the breaking down of food into simpler more usable forms.
    3 - respiration: the taking in of oxygen and breaking down of sugars to release needed energy.
    4 - excretion: the getting rid of waste material (what is not used or needed after digestion or respiration).

Volvox

A **precise structure** is now also considered a characteristic of living things. That means a living thing can usually be identified by the way it looks.

Robin

Scouring rush

NEEDS OF LIVING THINGS

Secondary consumer

Producer

All living things have some basic needs that must be met in order for them to survive. An understanding of these six basic need will help develop an understanding of how and why living things function as they do.

1 - ENERGY is needed for all kinds of activities such as moving and growing. The main source of energy in the world is the sun. Plants change the light energy to chemical energy (sugars) which they use for their own present needs or store for future use. This stored energy is one of the reasons animals eat plants.

Because plants make energy usable, they are called **producers**. An animal that eats plants or other animals is a **consumer**. An animal that eats only plants is said to be a **primary consumer** and an animal that eats other animals is a **secondary consumer**. Organisms that get their energy from breaking down producers or consumers are **decomposers**.

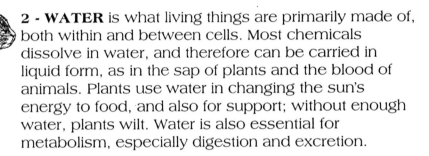
Primary consumer

Decomposer

2 - WATER is what living things are primarily made of, both within and between cells. Most chemicals dissolve in water, and therefore can be carried in liquid form, as in the sap of plants and the blood of animals. Plants use water in changing the sun's energy to food, and also for support; without enough water, plants wilt. Water is also essential for metabolism, especially digestion and excretion.

3 - OXYGEN is needed for respiration by almost all living things. It is taken out of the air with lungs, or else from the water with gills. Certain bacteria are **anaerobic**, which means they can live without oxygen.

4 - FOOD is anything that is eaten or absorbed for energy and nourishment.

5 - LIVING SPACE means more than just enough room to grow; it includes needed light, food and water. When these things are limited— and they always are limited to a degree—then there is **competition**. That means living things are struggling to get their needs met, and will often spend much time and energy defending what they have. This living space, especially for animals, is also called their **territory**.

Producer
Simplified food chain

Secondary consumer
Lacewing eats aphids

6 - PROPER TEMPERATURES are needed for most metabolic activities; when the temperature gets too hot or too cold, metabolism stops and the organism dies. Most living things have a particular range of temperatures in which they can live and have special features or behaviors that help them survive in it. For example, deciduous broadleaf trees drop their leaves before winter; some animals **hibernate** (sleep through the winter) while others **estivate** (sleep through the heat of summer days).

Animals that are warmblooded, such as birds and mammals, are not as affected by outside temperatures as are coldblooded animals, such as fish and reptiles.

THE CELL

Consider the toy building block. Although simple, it is the basic unit from which complex structures can be made. The basic block determines what the creation looks like and what it can do. With specialized blocks or attachments, new structures can be created which can function in new ways.

Building block

With the invention of microscopes, scientists were able to see that living things were all made of tiny units that looked like compartments. Therefore these units were called **cells** and were found to be the "building blocks" for all structures and functions of living things.

What became known as the **cell theory** states these three things:

1 - living things are all made up of cells.
2 - cells are the basic unit for structures and functions of all living things.
3 - all cells come from other living cells (not from dirt, or mud or nonliving matter as in the theory of spontaneous generation.)

There are things about cells that are important to understand for the purpose of **classifying** (organizing or grouping) living things. First of all, the outer covering of cells includes a thin, flexible **cell membrane**. In plant, fungi and bacteria cells, a stronger more rigid **cell wall** is outside of the cell membrane. Within the cell,

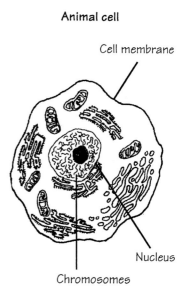

Animal cell

Cell membrane

Nucleus

Chromosomes

Plant cell

Cell wall
Cell membrane
Nucleus
Chromosomes
Chloroplast

acting like the control center or brain, is the **nucleus** (plural is **nuclei**). This round or oval structure has within it a thread-like mass; each "thread" is called a **chromosome**. All living cells have chromosomes in their nucleus except the single-celled organisms known as monerans which include bacteria. The chromosomes in these organisms spread within the center of the cell.

Plant cells are different from most other cells in that they have structures called **chloroplasts**, which contain **chlorophyll**. This is a colored substance or **pigment**, that uses the sun's energy to make food. Animal cells and fungi cells have no chlorophyll at all, while some protists and bacteria do.

CLASSIFICATION

With the incredible variety of living things, we need to have a system for grouping or organizing them. Our minds need to have a way to see relationships among them all. A classification system has been developed based upon the cell structure and whether the organism produces, absorbs or ingests food.

THE FIVE KINGDOMS

MONERA: cell wall
no nucleus

PLANT: cell wall
nucleus
chlorophyll
makes food
(photosynthesis)

FUNGI: cell wall
nucleus
no chlorophyll
absorbs food

PROTOZOA: no cell wall
nucleus
no chlorophyll
single-celled organism
makes, absorbs
or ingests food

ANIMAL: no cell wall
nucleus
no chlorophyll
many-celled organism
ingests food

There used to be only two main groups, called **kingdoms**: a living thing was considered either a plant or an animal. As the chart on the previous page shows, there are now five kingdoms, and even this may change as more is learned. The system of classification actually involves seven levels beginning with the most general similarities in the cells. They are:

    1 - **Kingdom**: main groups
    2 - **Phylum**: divisions of kingdoms
    3 - **Class**: divisions of phyla
    4 - **Order**: divisions of classes
    5 - **Family**: divisions of orders
    6 - **Genus**: divisions of families
    7 - **Species**: divisions of genera

The members of a species can even be divided into subspecies, varieties and breeds. The science of classifying is called **taxonomy**; it can become very complex, with levels in between the main seven.

Scientists have also developed a system for naming individual organisms. This **nomenclature**, as it is called, uses the genus and the species as the scientific name. Common names can vary from place to place and can be confusing, but scientific names are the same everywhere and in every language. Two ancient languages, Latin and Greek, are used for scientific nomenclature. (See Insect Orders chart page 25 for names and English translations.)

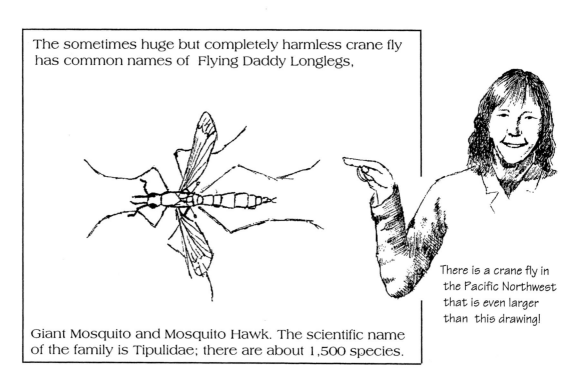

The sometimes huge but completely harmless crane fly has common names of Flying Daddy Longlegs, Giant Mosquito and Mosquito Hawk. The scientific name of the family is Tipulidae; there are about 1,500 species.

There is a crane fly in the Pacific Northwest that is even larger than this drawing!

Musca domestica
(housefly)

Kingdom: *Animalia*
Phylum: *Arthropoda*
Class: *Insecta*
Order: *Diptera*
Family: *Muscidae*
Genus: *Musca*
Species: *domestica*

Scientists classify an organism by its similarities and differences in structure compared to other living things. The common housefly is classified as follows:

1 - **Kingdom** - It is multicellular, and its cells have a nucleus, but no cell wall or chlorophyll; therefore it is classified as an **animal**.
2 - **Phylum** - It has jointed legs and a hard outer shell, which are the characteristics of **arthropods**.
3 - **Class** - It has only six legs, three body parts, two antennae and two eyes, which are the characteristics of **insects**.
4 - **Order** - It has only two wings instead of four, so it is a **diptera** (meaning two-winged).
5 - **Family** - Based on other characteristics such as veins in the wings, this fly is a **muscid**.
6 - **Genus** - This particular muscid is in its own genus, ***Musca***.
7 - **Species** - It is in the species **domestica** because it is the common domestic or housefly. The name is simply the genus (capitalized) and the species (not capitalized). ***Musca domestica*** is either written in italics or underlined.

LEVELS OF ORGANIZATION

Another system of classification involves relationships of working together. There are five levels of organization. They are from the simplest to the most complex:

1 - **Single Cell**
2 - **Tissues**
3 - **Organs**
4 - **Organ System**
5 - **Organism**

SINGLE CELL organisms include monerans, protists, some fungi and many simple plants. At this first level, the single cell lives and functions by itself.

TISSUES are cells that are similar, work together and do the same kinds of things. Muscle tissue is an example.

All Living Things

ORGANS are different kinds of tissues working together, each doing a particular job. The heart, for example, has muscle tissue, nerve tissue and blood tissue all doing separate things but working together to pump blood.

ORGAN SYSTEMS are groups of organs that work together. The heart, along with the blood, veins, arteries and capillaries make up the circulatory system which transports digested food, oxygen and wastes throughout the body. Other organ systems are: skeletal, muscular, digestive, excretory, respiratory, nervous, reproductive, endocrine (hormones) and even skin.

ORGANISMS are living things. A bacteria is an organism even though it is only one cell because it can live independently. A human is an organism at the more complicated fifth level: single cells, tissues, organs and organ systems are all working together.

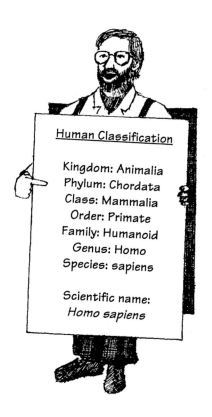

Human Classification

Kingdom: Animalia
Phylum: Chordata
Class: Mammalia
Order: Primate
Family: Humanoid
Genus: Homo
Species: sapiens

Scientific name:
Homo sapiens

INVERTEBRATES
(to the tune of "Clementine")

Traditional
Lyrics by Doug Eldon

Many creatures have different features... Yet all have a common trait... with no backbone they are all known to be called in-ver-te-ate.

All invertebrates together make a bit
 over 89 percent
Of the animal species in the world
 They live in any habitat

They're divided into phyla,
 by their structures classified
Here are eight kinds if you do find
 can then be identified

Poriferans are really sponges
 and they all have tiny pores
Cells in colonies they surely are pleased
 living on the ocean floor

Cnidarians like jellyfish, coral
 hydra, sea anemone
Cells for stinging have one opening
 hollow body cavity

Platyhelminths are the flatworms
 like the small planarians
Nematodes are little roundworms
 segmented worms are annelids

Echinoderms have spiny skins and
 tube feet coming out of them
Sea cucumber, starfish, sand dollar
 and the spiny sea urchin

Slugs and snails are one-shelled mollusks
 clams and scallops have two shells
Octopus, squid, nautilus are
 headfooted with tentacles

They all have a softish body
 with a mantle that can make
A hard shell that you can tell will
 give protection for their sake

Arthropods have jointed legs
 and a hard exoskeleton
Centipedes, millipedes, insects,
 arachnids, crustaceans

Centipedes have fewer legs and
 they are also carnivores
Millipedes have many more legs
 scavengers and herbivores

Crabs and lobsters, shrimp and barnacles
 are crustaceans you can tell
Four antennae legs are many
 they're aquatic with a shell

Arachnids include the spiders
 Ticks and mites and scorpions
They all have eight legs two body parts
 No antennae and no wings

Insects have six legs, three body parts
 two antennae and two eyes
Egg to larva change to pupa
 then to adult with wings to fly

INVERTEBRATES

Hydra

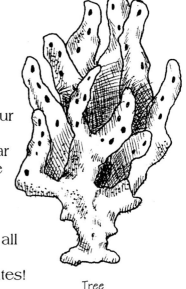
Tree sponge

The kingdom of animals is divided into thirty-four subgroups, called **phyla**, based on body structures. Thirty-three of these phyla are similar in one way; they do not have backbones made of vertebrae. Because they lack vertebrae, all these animals are called **invertebrates**.

Only eight phyla will be considered here, but if all the different species of animals were counted, almost 90 percent of them would be invertebrates!

1- PORIFERANS, or sponges, live in fresh salt water, and are very simple organisms. The individual cells of poriferans live together in a mass called a **colony**. But a fragment of a porifera colony is able to regrow, or **regenerate** an entire new body! Each kind of sponge has a different shape, but they all have pores through which water passes, this is how the phylum was named porifera. The sponges we now use to wipe things with are man-made, but real sponges can still be found in some stores.

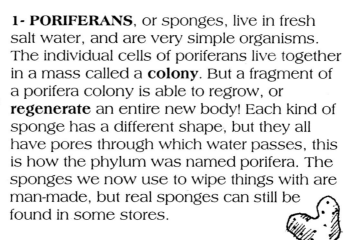
Sea anemone

2 - CNIDARIANS were known as **Coelenterates** until recently. They all have a hollow body structure with one opening, and tentacles with stinging cells. These specialized cells that stun, paralyze or kill their prey are known as **nematocysts**.

Hydras are the simplest in this group. **Corals** are individual organisms that live in colonies and build hard coverings. Over time, masses of these colonies build up to form a reef. **Sea anemones** look somewhat like flowers (there is a flower called an anemone) and live in tidepools and reefs.

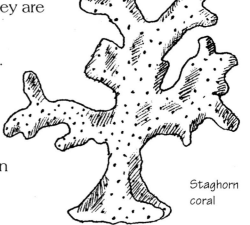
Staghorn coral

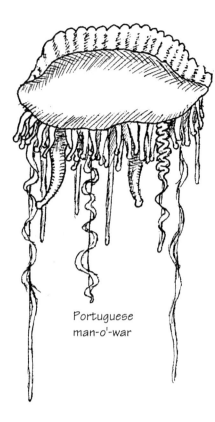

Portuguese man-o'-war

Jellyfish come in many shapes, sizes and colors. Some can be very dangerous, such as the Portuguese man-o'-war. The Australian sea wasp, which is really not a true jellyfish, can kill a person in 1-3 minutes![1]

3 - **ECHINODERMS** have spiny skins (*echino* means spiny, *derm* means skin) and little suction-tipped tube feet. The phylum is divided into several classes, all of which live in the ocean. **Sea stars**, also called **starfish**, usually have bumps instead of spines; have from 5 to 20 or more arms (which they can regenerate if broken off) and have tube feet in rows on the underside of the arms. **Sea urchins** (named after the spiny hedgehogs called urchins) have spines that can be quite long, like a pin cushion. **Sand dollars** are somewhat flat, with spines that look more like fuzz that rubs off easily when the animal dies. **Sea cucumbers** do look like that garden vegetable and **sea lilies** look more like flowers than animals.

There are at least 16 phyla that are called worms. Here are three considered quite important:

4 - **PLATYHELMINTHS** are also called **flatworms** (*platy* means flat, like a plate, and *helminth* means worm). A common example is ***Planaria***, often used in biology experiments because it will grow two whole bodies if it is cut in half. Tapeworms are another kind of flatworm. They often live in warmblooded animals, including man. Some of these live in intestines and grow up to an inch wide and 20 to 50 feet long!

5 - **NEMATODES** are worms with a round body that is usually threadlike with pointed ends. Most of these **roundworms** are microscopic and harmless, eating mainly bacteria, protozoa and fungi. Some are **parasitic**, meaning they are harmful to the organism they live on or in. They may be parasitic on plants, causing losses in food crops, or in animals causing health problems.

One of the most serious is the **hookworm**. This parasitic roundworm infects the intestines of hundreds of millions of people, especially in Africa, by burrowing through and entering bare feet.

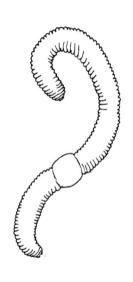

Earthworm

Planaria

Invertebrates

6 - ANNELIDS are different from most other worms in that their bodies are divided by rings into segments or sections. **Earthworms** and **leeches** are examples, but there are many other kinds of segmented worms, especially in the ocean.

Tower shell: a kind of snail

Chambered nautilus

7 - MOLLUSKS all have a soft body, are often covered with a hard shell and usually have a thick muscular foot. The hard shells that protect the soft body parts of mollusks are produced by materials made by the **mantle**, a tissue which covers the body. The phylum is divided into eight classes by characteristics such as type of foot, type of shell and presence of a shell.

Univalves are mollusks with only one shell, or no shell at all. These are also called **Gastropods** which means stomach-foot. This group includes slugs and snails that live on land and in water.

Slug

Bivalves, or **pelecypods** are mollusks with two shells, such as clams, scallops, mussels and oysters.

Cephalopods (*cephlalo* means head, *pod* means foot) all look like a large head attached to a group of leg-like tentacles. This group includes the chambered nautilus which has a shell; the octopus which has eight tentacles; and the squid and cuttlefish, which have ten tentacles, two of which are longer and somewhat paddle-shaped. The Atlantic giant squid can grow to over 50 feet long, and is the world's largest invertebrate.[2] It also has the largest eye: over 15 inches in diameter![3]

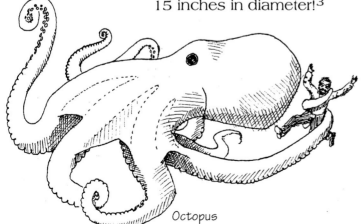
Octopus

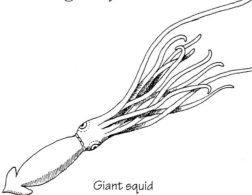
Giant squid

8 - ARTHROPODS are a very large group: there are more different kinds of arthropods than all the other kinds of animals combined! One characteristic of animals in this phylum are legs that have joints and bend much like your elbow or knee. In fact, the name comes from *arthro* which means jointed, and *pod* which means leg. The other characteristic of these animals is a hard outer covering or shell—a skeleton on the outside of their bodies called an **exoskeleton**. The phylum Arthropoda is divided into five classes, based on the number of body parts such as legs, antennae, and eyes. They are:

  1 - **Crustaceans**
  2 - **Chilopoda (Centipedes)**
  3 - **Diplopoda (Millipedes)**
  4 - **Arachnids**
  5 - **Insects**

Crustaceans have many body segments and many legs—at least ten. They often have claws or pinchers and four antennae; because almost all live in water, they also have gills for breathing.

Many crustaceans are harvested as seafood, including crabs, lobsters, shrimp, prawns and freshwater crayfish (crawdads). Barnacles are commonly attached to rocks, fish or other objects in the ocean, including the hulls of boats. There are also many small crustaceans that live in pond water, such as **Daphnia** (water fleas). Some crustaceans are **terrestrial**—they live on land—and have legs that all look the same. They are the **isopods**; they include pill bugs, which can roll into a ball, and sow bugs which cannot.

Centipedes and millipedes used to be in the class **Myriapoda**, which means many-legged. They are actually different from each other in several ways.

Invertebrates

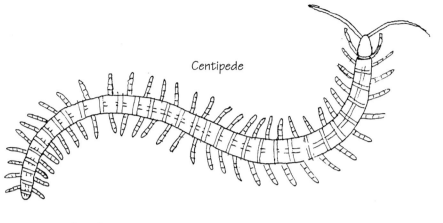
Centipede

Millipede

Centipedes have one pair of legs on each body segment, are somewhat flat, wiggle from side to side like a snake, and are active **carnivores** (eat animals, especially insects). Millipedes have two pairs of legs on each body segment, are more cylindrical, do not wiggle like a snake but can coil up, and are **herbivores** (eat plants) and/or **scavengers** (eat whatever they can find). Centipedes are now their own class, **Chilopoda**, and millipedes are also in their own class, **Diplopoda**.

Arachnids include spiders, harvestmen (daddy longlegs, which are not true spiders), scorpions, ticks, mites, horseshoe crabs and sea spiders. They are different from other arthropods in that they all have eight legs and two body parts. They do not have antennae, wings or compound eyes as insects do, but may have many simple eyes, or no visible eyes at all. Most people are familiar with spiders, especially their feeding habits, and scorpions with their pinchers and stinger-tails. Ticks attach themselves to mammals and suck their blood. Mites are similar, but live on plants or animals and can be very, very small. One kind of mite that lives in the trachea (windpipe) of honeybees has become a serious problem, killing off thousands of hives.

Tick

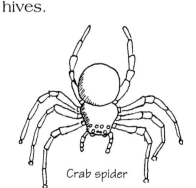
Crab spider

Scorpion

Direct metamorphosis

Silverfish

Insects are by far the largest group, not only of arthropods but of all animals! The class **Insecta** includes so many different kinds that it is divided into more than twenty-five orders. One of those orders, **Coleoptera** (beetles), has more than 200,000 different species!

Insects have only six jointed legs and a body that is divided into three sections—**head, thorax**, and **abdomen**. The head has two antennae, usually two main eyes that are **compound** (made up of many eyes) and a mouth that may be for chewing, sucking or lapping. The thorax is where the legs and wings of most adult insects are attached. (Insects are the only invertebrates that can fly.) The abdomen normally has eleven segments but they are not always easy to see. Insects are considered the most successful animals: they have adapted to live in almost every habitat on land, in water and in the air.

METAMORPHOSIS

Growth and development, or **metamorphosis**, is unique in insects. There are three kinds:

Incomplete metamorphosis

Dragonfly larva (naiad)

1 - **Direct**: the egg hatches to a stage that looks like an adult, but is smaller and not mature.

2 - **Incomplete, simple** or **gradual**: the egg hatches into a small larva which may look different from the adult. The larva at this stage is called a **nymph**, or if aquatic, a **naiad**.

Complete metamorphosis

1-Caterpillar

3 - **Complete** or **complex**: the egg hatches into a **larva** (which usually eats a lot) then changes into a **pupa** (which usually neither eats nor moves), and finally hatches into an **adult**. An example of this is a moth: an egg hatches into a caterpillar (larva), which turns into a cocoon (pupa) from which an adult emerges.

2-Pupa

3-Adult Moth

CHARACTERISTICS OF COMMON INSECT ORDERS

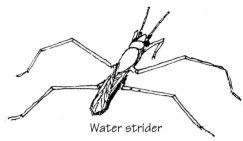
Water strider

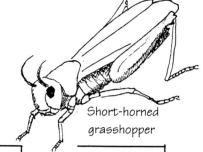
Short-horned grasshopper

Order ("ptera" = wing)	Example	Metamorphosis	Importance
Isoptera "same winged"	Termite	Incomplete	Destroy wood buildings
Odonata "toothed"	Dragonfly Damsel Fly	Incomplete	Eat harmful insects
Orthoptera "straight winged"	Grasshopper Cockroach	Incomplete	Damage crops
Hemiptera "half winged"	All True Bugs Water Strider	Incomplete	Damage crops Carry disease
Homoptera "like winged"	Aphid Mealy Bug	Incomplete	Damage crops
Neuroptera "nerve winged"	Lacewing Dobson Fly	Complete	Destroy harmful insects
Diptera "two winged"	Fly Mosquito	Complete	Carry disease Act as pests
Lepidoptera "scale winged"	Butterfly Moth	Complete	Pollinate flowers Produce silk Damage crops
Coleoptera "sheath winged"	Ground Beetle Weevil	Complete	Damage crops Act as pests Eat other insects
Hymenoptera "membrane winged"	Bee Wasp Ant	Complete	Pollinate flowers Act as pests Make honey

Cockroach

COLDBLOODED VERTEBRATES
(to the tune of "When Johnny Comes Marching Home")

Patrick S. Gilmore

Lyrics by Doug Eldon

Oh, when you study animals, there're some of which you're told, whose blood will always stay quite warm; but some whose blood is cold. They can survive within a range of temperature except a change.. that's too extreme can be so dangerous for animals. For coldblooded vertebrates; coldblooded animals.

They must respond by what they do and so they move around
To find the proper temperature in water or on ground
Fish, reptiles, amphibians; whose temperature of blood has been
Controlled from outside not within these kinds of animals
These coldblooded vertebrates; coldblooded animals

The fish have gills instead of lungs to get their oxygen
And most have air bladders and fins to help them float and swim
Some are jawless, like lampreys with cartilage like sharks and rays
But most have bony vertebrae, they're bony animals
They're coldblooded vertebrates; coldblooded animals

Amphibians lead double lives, that's how they get their name
They start in water, then go on land, which they like just the same
Toads have dry and bumpy skin, 'frog's is wet and smooth as in
The salamanders and newts, their kin. They're all amphibians
They're coldblooded vertebrates; coldblooded animals

Snakes and lizards are reptiles, along with tortoises
Whose legs are made for walking, not for swimming like turtles
Crocodiles have teeth that show, an alligator's will always go
Inside its mouth that's how to know these different animals
These coldblooded vertebrates; coldblooded animals

    These animals are different in the way they reproduce
    The fish must fertilize their eggs externally and loose
    Amphibians' jelly-eggs are wet reptiles' are leathery and set
    To hatch on land where they will get to become animals
    These coldblooded animals; coldblooded animals

COLDBLOODED VERTEBRATES

Only one phylum of animals has the characteristic of a backbone made up of vertebrae. These animals are known as **vertebrates**, or **chordates** because they have a spinal cord. The phylum is divided into eleven groups called classes. Of those, nine are **coldblooded**, which means they depend on the temperature outside their bodies to get and keep them warm or cool. See appendix for classification of some common animals.

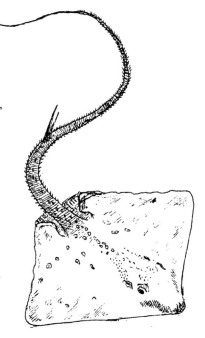
Stingray

Column of vertebrae

A snake or lizard may lie in the sun in the morning to get their blood warm. Then, in the heat of the day they find shade so their blood does not get too hot. That is the only way they can control their body temperature; their blood is as cold or as warm as the air or water around them.

Most coldblooded animals live within a range of temperatures where it does not get too hot or cold. For example, some fish live only where the water is cold, while others live only in warm, tropical waters. If a tropical fish is put in cold water, it would soon die—and a cold water fish would not survive in warm tropical water.

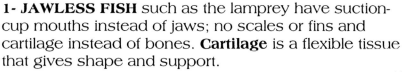
Lamprey

FISH

Vertebrates in the following three classes:
1 - breathe through **gills** for getting oxygen out of the water.
2 - have fins to help them swim, steer, and stop.
3 - reproduce with eggs being fertilized outside the female.

The end of your nose is made of flexible cartilage.

1- JAWLESS FISH such as the lamprey have suction-cup mouths instead of jaws; no scales or fins and cartilage instead of bones. **Cartilage** is a flexible tissue that gives shape and support.

2- CARTILAGINOUS FISH (car-til-a-jin-us) include sharks, rays and skates. They also have cartilage instead of bones and rough skin instead of scales; but they definitely have jaws! Most of them have fins, which in rays and skates look more like wings. In fact,

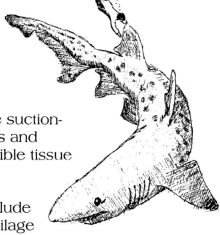
Shark

some rays can actually fly out of the water and glide through the air along the surface.

3- BONY FISH are all the other fish. In addition to a bony skeleton, they have jaws, fins, scales and **air bladders** to help control their floating. There are 24,000 different kinds living in fresh water, salt water, or both. Most of the fish we eat are in this group, including trout, salmon, perch, cod, tuna and sardines.

Bluegill

AMPHIBIANS

This class of vertebrates includes toads, frogs, newts and salamanders. All of them have life cycles that begin in water where the eggs are laid. The eggs of amphibians have a jelly-like coating and may be single, in a strand, or in a mass. The young that hatch from these jelly-eggs are called **tadpoles** or **polliwogs**. Frog tadpoles eat water plants such as algae; breathe through gills; have a long tail; but do not have legs. They gradually lose their tails and gills and grow lungs and legs. Adult frogs and toads live on land or in water and eat insects or other invertebrates. Salamanders and newts eat small aquatic insects instead of plants when young, and do not lose their tails.

Frog

Salamander life cycle

1- Eggs

Frogs, salamanders and newts could be considered "kin," or close relatives, because they are similar. All of them have smooth and moist skin through which they breath. (You should always have wet hands when handling them. The oil in your dry skin will clog the pores in the amphibians' skin, and they may suffocate.)

2- Polliwog

Toad

Salamanders and newts do not jump, but walk on land and have long tails to help them swim.

Frogs and toads look much the same, but the skin of toads is dry and bumpy; the skin of frogs is moist and smooth.

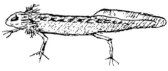
3- Eft

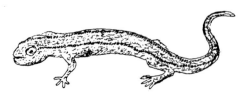
4- Adult

REPTILES

Reptiles are different from amphibians in several ways:

Lizard

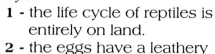
Rattlesnake

1 - the life cycle of reptiles is entirely on land.
2 - the eggs have a leathery shell, so they don't dry out or break.
3 - there is no tadpole stage, but young and adults look similar.
4 - the skin of the adult reptiles is dry and scaly.

Most reptiles are carnivores (meat eaters) and some can be dangerous.

Reptiles include:

1 - snakes.
2 - lizards.
3 - turtles, which live in water and have paddle-like legs.
4 - tortoises, which live on land and have stumpy legs for walking.
5 - crocodiles, which have narrow, pointed snouts and bottom teeth that show when the mouth is closed.
6 - alligators, which have wide snouts and bottom teeth that can not be seen when the mouth is closed.

Crocodile

Alligator

Not all animals that live in the water are coldblooded. Whales are not fish but mammals, which are warmblooded. That is why whales such as gray whales migrate from Alaska, where the water is very cold, to the Gulf of California, where the water is warm. Their bodies are able to keep the same temperature even when the water temperature changes. Marine mammals (including dolphins, seals, otters and walruses) breathe air with lungs; fish are different in that they get oxygen out of the water through their gills.

Tortoise

Turtle

BIRDS
(to the tune of "If You're Happy and You Know It")

Animals that people study quite a bit....
that are warmblooded and are also vertebrate......
We do classify as birds... now we'll share with you some words,
so some interesting facts you won't forget.

Characteristics are a beak, two legs and wings
Several kinds of feathers which do different things
There are some that help them fly
Others make them more streamlined
Fuzzy, down feathers are insulating

Bones of birds are light because they are hollow
So it's easier to fly, and don't you know—
Both to cool them when they're flying
And for oxygen supplying
They have air sacs to help them on the go

Eggs let oxygen pass right on through the shell
Often laid inside a nest constructed well
The eggs the parents incubate
Hatch their young and may migrate
Following their food supply—there's more to tell

The class of birds has many orders and families
Which are divided into genus and species
But nine thousand different kinds
Are too many for your minds
So forget about more verses, if you please

BIRDS
Our Fine Feathered Friends

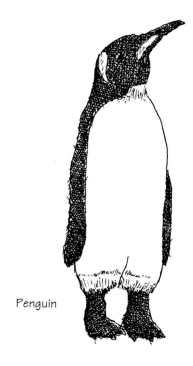
Penguin

Birds are one of the most popular animals, other than domestic animals such as dogs or cats. Many people make a hobby out of studying, watching or even just counting birds. Libraries and bookstores usually have more books about birds than any other kind of animal because there are people everywhere who are simply fascinated with birds.

Down feathers

CHARACTERISTICS

Birds are all warmblooded vertebrates. They have two legs, two wings, a beak, feathers that cover the body and young that hatch from an egg.

There are four kinds of feathers, but two are most important. **Contour** feathers on the wings help birds fly; they also cover their body and make them more streamlined. **Down** feathers are for **insulation**, keeping the bird warm. They cover young birds and are found on the breasts of adults. People use the down in jackets and sleeping bags because it is light weight and keeps body heat in and cold out.

Birds have **hollow bones** which make flight easier because they are light. **Air sacs** are hollow structures connected to the birds' lungs. As birds fly, these air sacs help provide a constant supply of oxygen to the lungs and act somewhat like an air cooler for the birds' insides.

Contour feather

Air sac system branches off the lungs

Eggs of birds are similar in several ways to plant seeds: **1)** they both contain a living **embryo**; **2)** they have a supply of food for the developing young and **3)** they both are protected by a hard outer shell. This shell allows oxygen and carbon dioxide to pass through to the embryo.

BEHAVIOR

Nests are built by most birds before any eggs are laid. The nest may be quite simple or they may be very complex. A killdeer often makes a nest among rocks and lays its mottled eggs right on the ground—but both the bird and the eggs can be hard to see! Most birds build nests up higher, constructing them out of various materials, to provide protection and a home for the eggs and the baby birds. Nests can be as small as a thimble (that of a bee hummingbird) or as big as a bedroom (that of an eagle).

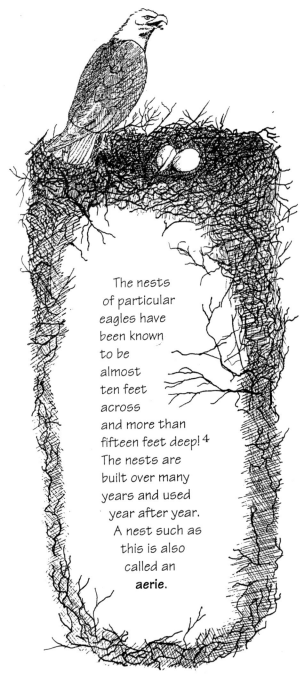

The nests of particular eagles have been known to be almost ten feet across and more than fifteen feet deep![4] The nests are built over many years and used year after year. A nest such as this is also called an **aerie**.

For the eggs to hatch, they must be kept at the right temperature. Parents **incubate** the eggs; sitting on them to keep them warm. After they hatch, the parents supply the young birds with food and teach them the skills they need for life on their own.

Birds **migrate** when they move to a new environment, most often to find food. Migration usually involves flying to where the climate is warmer. Some fly several hundred miles, but others fly many thousands—the arctic tern migrates up to 14,000 miles!

Birds have very strong **instincts**; without being taught, they know when and how to build nests, and when and where to migrate. But they do learn and actually may be much smarter than we think. Crows have quite a reputation for giving farmers a hard time and geese are sometimes used to keep watch over other animals, as a guard dog would. Many parrot-like birds talk and can even carry on a conversation. One knew nearly 800 words![5]

Birds

CLASSIFICATION

Birds are in the class of warmblooded vertebrate animals called **Aves**, of which there are about 9,000 species. They are generally grouped according to where they live, what they eat and where they rest.

Mallard duck

Land Birds are most of the perching birds and the songbirds familiar in forests, meadows and backyards:

    Blackbird   Robin   Woodpecker
    Blue Jay   Sparrow   Pigeon
    Crow   Swallow   Starling

Game Birds are those hunted for sport or food:

    Duck   Pheasant
    Goose   Quail

Passenger pigeon

Water, Marsh, and Wading Birds refer to those that live on or near fresh water, such as lakes and marshes:

    Heron   Stork
    Loon   Swan

Oceanic and Beach Birds live on or near salt water:

    Albatross   Pelican
    Gull   Tern

Parrots and Tropical Birds include many kept as pets and those with colorful feathers:

    Flamingo   Parrot
    Parakeet   Macaw

Flightless Birds include those that depend on their legs for running or their wings for swimming:

    Kiwi   Penguin
    Ostrich   Rhea

Birds of Prey are strong fliers with keen eyesight, sharp **talons** (or claws) and strong curved beaks—all for catching and eating animals:

    Eagle   Osprey
    Falcon   Owl
    Hawk   Vulture

Extinct Birds lived long ago and died out on their own or were killed off by humans:

    Dodo   Passenger Pigeon

Heron

PLANTS
An Introduction

The kingdom of plants is often divided into two large subkingdoms: those which are simple, which do not have vascular tissue (**nonvascular** plants), and those which are more complex and have vascular tissue (**vascular** plants).

VASCULAR TISSUE

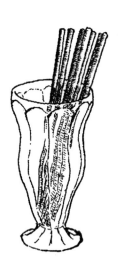

Imagine holding a bunch of drinking straws. If you were to put one end of this bundle of straws in a liquid, and sucked at the top end, you could **transport** (move) the liquid up into your mouth. Those straw bundles would be working like the specialized cells that are organized into transporting tubes in plants. The word **vascular** simply means vessels for carrying. Veins and arteries in animals are vessels for carrying blood, and are part of the vascular system. Plants have veins and vascular tissue for carrying **sap**, which is the "blood" of plants.

Vascular plants can draw water and the nutrients out of the ground and transport them, against the pull of gravity, up to the top leaves which may be hundreds of feet high. The food that is made in the leaves is transported down through the vascular tissues to the lower parts of the plant.

Nonvascular plants must live in a moist environment—in or near water—because they do not have a way of transporting water, nutrients or food great distances.

REPRODUCTION

Another way the kingdom of plants is divided is into those that reproduce by spores and those that reproduce by seeds. Most vascular plants reproduce by seeds but ferns, horsetails and a few others are exceptions—they are vascular plants that reproduce by spores. Most of the nonvascular plants do not have seeds or anything involved in making or protecting seeds, such as flowers, fruits or cones. Instead,

Bull kelp: brown algae

they have tiny reproductive cells called **spores**. Nonvascular, spore-producing plants include algae, moss and liverworts.

CLASSIFICATION

Classification systems are often changing. Instead of living things being considered either plant or animal, now they may be grouped in one of five kingdoms, as discussed earlier on page 14. For example, scientists are now grouping algae with protists instead of plants. This is because the cells of algae are not specialized as most plant cells are. Large algae such as seaweed are now considered to be more like a large colony of single-celled organisms than a multicellular plant.

Fungi (or fungus) are not considered plants because the cells do not have what plant cells have for making food. True plants have a substance known as **chlorophyll** in their cells which uses the energy from light to make food (sugars) out of water and air. Fungi can't make food; the needed energy is absorbed from living or dead matter. Because of these differences, fungi are now their own kingdom. But since fungi are nonvascular and they reproduce by spores, they are sometimes included when plants with these characteristics are being discussed.

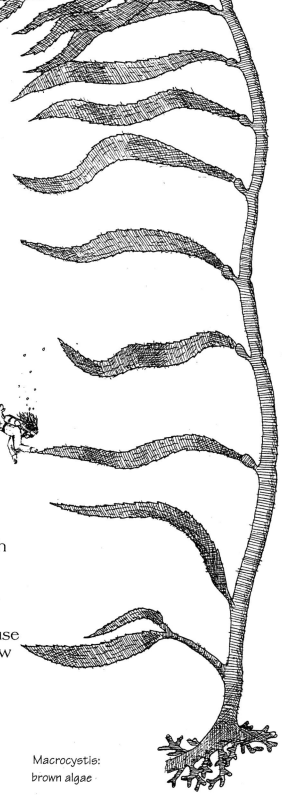

Macrocystis: brown algae

ALGAE, FUNGI
AND
NONVASCULAR PLANTS
(to the tune of "Go Tell Aunt Rhodie")

Al- gae and fun- gi, li- chen, moss and liv- er- worts all are non- vas- cu- lar and re- pro- duce by spores. Al- gae is cla- si- fied by co- lor in- to fi- ve groups. They can be green, or brown, gold- en, red, or fire. Fun- gi lack chlo- ro- phyll, they get energy other ways. Most by de- com- pos- ing, or fer- men- ta- tion. If they live on dead things, they are known as saprophytes. If they feed on living things, then they're par- a- sites.

Mushrooms and toadstools, molds and mildews, yeast and rots
Are many kinds of fungi: some are good, some not
Lichen's really two things, living symbiotically
(Helping each other): algae and fungi

Mosses and liverworts, found in moist environments
Are simple, green, nonvascular spore-producing plants

ALGAE, FUNGI AND NONVASCULAR PLANTS

Different kinds of chlorophyll make plants different colors. Algae is classified by the kinds of chlorophyll in the cells or, more simply, by the color of the algae.

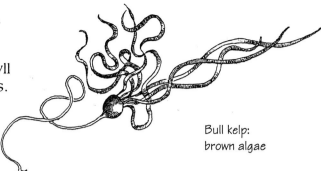
Bull kelp: brown algae

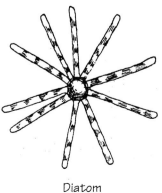
Diatom

Green algae includes many that are only single-celled, others that are long threads of cells and still others that are many-celled and leaf-like. Green algae grow well in ponds, lakes and streams, especially if the water is warm, and even when it is polluted. Swimming pools will grow green algae quite well if they don't have chlorine added!

Brown algae includes most seaweeds, especially the largest and most complex algae; the kelps. Some kelps grow up to eighteen inches a day, and may reach nearly two hundred feet![6] Kelp beds are somewhat like underwater forests of giant algae, which support all kinds of other marine life.

Spirogyra: green algae

Diatom

Red algae includes other seaweeds, some of which grow much deeper in the oceans than brown or green seaweeds. One kind was found over 800 feet underwater![7]

Golden algae includes **diatoms**: microscopic, free-floating algae that have cell walls made of a glass-like substance called **silica**. It is formed in fascinating shapes and reflects light to give them their golden color.

Diatom

Diatom

Fire algae are well known for two things: the **red tide** and **bioluminescence**. The red tide is caused by poisons produced from large blooms of some kinds of these free-floating algae. Bioluminescence is a kind of glowing produced by other kinds of fire algae. This glowing can occasionally be seen at night in water that has been stirred up by waves or boats.

As mentioned, algae are now considered protists instead of plants. Fire algae especially look and act in a way that make them more animal-like than plant-like. Another organism, **blue-green algae**, is no longer grouped with the other algae, but is in the kingdom Monera along with bacteria.

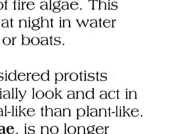
Diatom

FUNGI

Fungi are not really plants because they have no chlorophyll, but they reproduce by spores and are nonvascular.

Mushroom development — Cap, Gills and Spores, Ring, Stalk. Many strands of hyphae form a net of mycelium

Life Cycle: A spore from a fungus **germinates** (begins to grow) and forms tiny thread-like **hyphae**. The hyphae continue to grow and branch, forming a mass of white **mycelium**. Under certain conditions some of the hyphae in this mass begin to develop into a spore producing structure, also called a **fruiting body**. A mushroom is a kind of fruiting body—it is neither a fruit nor a vegetable. When these fruiting bodies are mature they release spores and the cycle begins again. Some fungi live for years before or after forming fruiting bodies. In fact, the largest and the oldest living thing in the world is thought to be a fungus that covers more than thirty-five acres and is thousands of years old!

Saprophytes are organisms that live on dead things. Examples of saprophytic fungi are **dry rot** and **wet rot** which **decompose** (break down) wood. These rots are important in forests or in a compost pile because they break down dead plants and animals and return nutrients to the soil. But these rots are certainly not appreciated when they break down the wood in a house or bridge!

Dry rot

Parasites are organisms that feed on living things. An example of a parasitic fungus is the kind that grows between toes: athlete's foot. There are many serious parasitic fungi that attack plants, including **powdery mildew, sooty mold, brown rot** (on fruit), and **root rot**.

Algae, Fungi and Nonvascular Plants

There are many sizes, shapes and kinds of fungi, some of which are quite important to us:

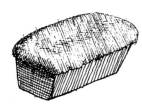

Rots decompose things, as mentioned.

Yeasts are important because they get their energy from a process called **fermentation**. As a by-product, yeasts produce carbon dioxide which makes bubbles in bread dough, causing it to rise; or in beer or champagne, making it fizzy. Another by-product is alcohol, which turns grape or other fruit juice into wine.

Molds can spoil food, such as bread or fruit, but can also give food a special flavor, such as in blue cheese.

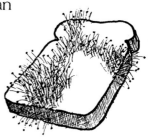
Common bread mold

Mildews are much like molds, but are seldom considered beneficial. They grow on just about anything that is damp, such as wet socks left in the bottom of your closet all winter. Mildews give off a musty smell, as in an old house that has been closed up for too long.

Bracket fungi

Mushrooms can be good to eat, and there are now many kinds to choose from in grocery stores. There are a variety of edible wild mushrooms too, but there are also many that can be very poisonous! It can be difficult to tell which are good and which aren't, so it is best not to eat wild mushrooms unless an expert is positive they are edible. The mushrooms you should not eat are often called **toadstools**; some are quite deadly.

SYMBIOSIS

Staghorn fungi

When two organisms live together in a relationship where both get something to their advantage, and neither is harmed in any way, they are said to be living in a **symbiotic** relationship called **mutualism**. Symbiotic simply means together-life and mutual means both-sharing.

Mushroom

One example of mutualism is between certain fungi and plants. The fungi wrap their threadlike hyphae around the roots of the plant to help it absorb water and nutrients out of the soil. The plant returns the favor by sharing with the fungus the food it has made. Each gets something it needs and neither is harmed. This particular mutualism is called **mycorrhiza**, which means fungus-root.

Cup fungi

Old man's beard lichen or goat's beard lichen

LICHEN

Lichen is a perfect example of mutualism; it is always an **alga** (one kind of algae) and a **fungus** (one kind of fungi) that live together and help each other. The alga has chlorophyll and so can make food. The fungus absorbs water and nutrients from whatever the lichen is attached to and makes sure it is attached well. In this way, lichen can grow where no other plants can grow, and where neither the alga nor the fungus could survive alone. Because lichens can grow where most plants cannot, such as on rocks, and because they are often the first to settle such areas, they are sometimes thought of as "pioneers."

Lichen can be many different sizes, shapes and colors. It can look like a colored crust growing in a patch on a rock, like a gray beard hanging from a tree branch or like a green leafy flap on a tree trunk.

MOSS

Mosses do not have true leaves, stems or roots, but they have structures that function like plant parts and are usually green. Becaus mosses do not have vascular tissue and cannot get very tall, they must live near sources of water. They often grow on trees, rocks, or directly on the ground. They also grow on roofs, especially on the side facing north which gets less sun and stays wet longer. Mosses dry up until the cool wet weather of autumn brings them back to life. Mosses sometimes grow in lawns, making them look healthy, but actually they choke the grass.

Moss

Bogs are masses of mosses, sometimes many feet deep. Peat moss is a common type used in potting soil and sphagnum moss is often used in flower arrangements.

LIVERWORTS

Liverworts are nonvascular plants that look like nothing but a little green leaf, usually growing near moss. Like mosses, they don't have true leaves, stems or roots but rather leaflike structures that sometimes are shaped like tiny livers.

Liverwort

Algae, Fungi and Nonvascular Plants

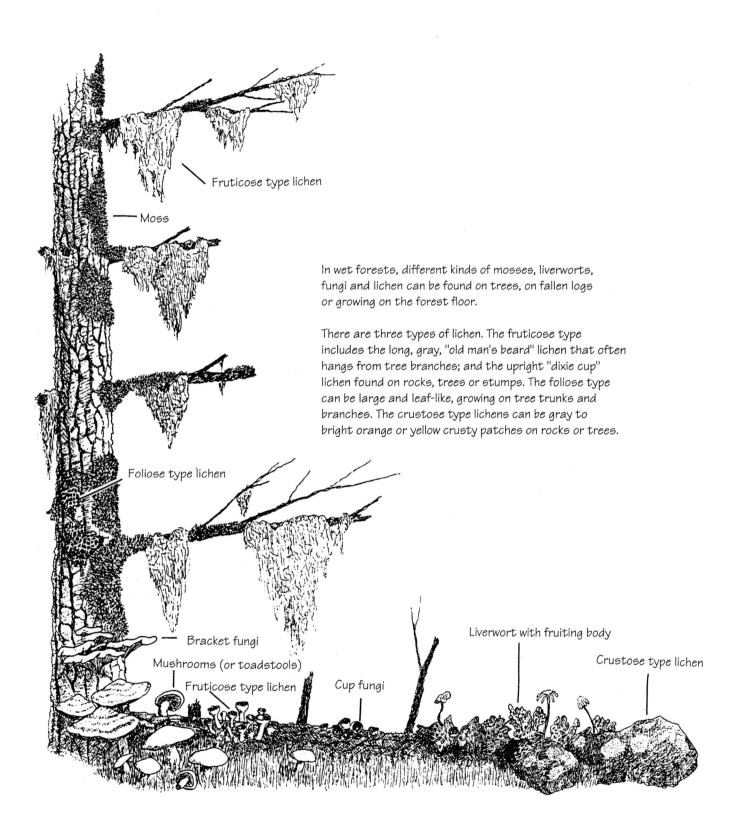

In wet forests, different kinds of mosses, liverworts, fungi and lichen can be found on trees, on fallen logs or growing on the forest floor.

There are three types of lichen. The fruticose type includes the long, gray, "old man's beard" lichen that often hangs from tree branches; and the upright "dixie cup" lichen found on rocks, trees or stumps. The foliose type can be large and leaf-like, growing on tree trunks and branches. The crustose type lichens can be gray to bright orange or yellow crusty patches on rocks or trees.

VASCULAR PLANTS
(to the tune of "Battle Hymn of the Republic")

William Steffe
Lyrics by Doug Eldon

Xy-lem car-ries min-er-als and wa-ter toward the sky.
Phlo-em car-ries food on down and that's the rea-son why. The most im-por-tant cells are those pro-duc-ing both of them, the cells of the cam-bi-um.

Chorus
Vas-cu-lar..... Oh vas-cu-lar plants.
Vas-cu-lar Oh..... vas-cu-lar plants.
Vas-cu-lar... Oh.... vas-cu-lar plants all have trans-port-ing tubes.

Ferns and also horsetails are both vascular indeed
They reproduce by means of spores and rhizomes, not by seed
The spores come from the sorus underneath the fronds of ferns
Growing into a prothallium

Gymnosperms have unprotected seeds which you may see
On the cones of conifers most often in a tree
Fir, pine, hemlock, cedar, spruce, redwood and juniper
Stay green throughout the year

Gymnosperms with naked seeds and angiosperms with fruits
Absorb the water from the soil and held in place by roots
Food is made in leaves and then the stems help to transport
That food throughout the plant which they support

Flowers may have sepals, petals, pistils, and stamen
Ripened ovaries are fruits with seeds inside of them
Seeds have coats and embryos that are indeed alive
With food to help them survive

CO_2 and H_2O with light for energy
Glucose is made by chlorophyll and oxygen's set free
(Respiration is the opposite of this process)
It's called photosynthesis

Plants need water, light and air with proper temperature
Space to grow, and minerals, they need them to mature
Tropisms are responses to stimuli, I'm told
By hormones they're controlled

VASCULAR PLANTS

Fern fronds
Sori
Rhizome
Fiddlehead
Roots

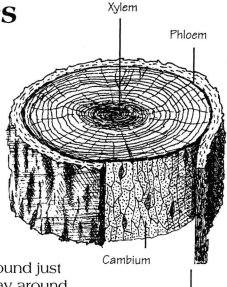

Xylem
Phloem
Cambium
Bark

VASCULAR TISSUE
The three kinds of vascular tissue are:

1 - xylem (zy-lem) carries water and minerals from the roots to the upper parts of the plant.

2 - phloem (flow-em) carries food from the leaves, where it is produced, down to lower parts such as the roots.

3 - cambium produces the cells of both the xylem and the phloem and is usually found just under the bark of trees. If it is cut all the way around, new xylem or phloem cells cannot be produced and the tree will die. This is called **girdling**, and bears have been known to do this to trees when sharpening their claws.

CLASSIFICATION

Vascular plants are classified according to how they reproduce, the structures involved in reproduction and by other structures such as leaves. Specific characteristics used for classifying include size, shape, color and number of parts.

The subkingdom of vascular plants is then divided into several **subgroups** called divisions (rather than "phyla' as in the animal kingdom). Two of these divisions are ferns and horsetails which have neither flowers nor seeds. These divisions reproduce by **spores** and by underground root-like stems called **rhizomes**.

One way ferns are classified is by their spores and the structures that the spores come out of, called **sori** (or sorus). These sori are generally on the underside of the leaf-like **frond**. When a fern spore begins to grow, it starts out as a tiny heart-shaped **prothallium**, which produces egg and sperm. If fertilized, the egg will then develop into a fern.

2- spores
3-prothallium
1-fern
Life cycle of the fern

Horsetail rush

GYMNOSPERMS

Another major plant division is made up of vascular plants which produce unprotected seeds. The division is called **gymnosperm** which is Greek for naked-seeds. The most common gymnosperms are **conifers** whose seeds are between the scale-like bracts of cones. Most conifers have needle-like leaves which stay green all year, therefore they are called **evergreens**.

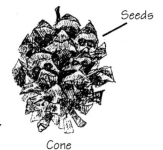

Conifers are classified using characteristics of needles, cones, bark, as well as the general shape of the tree and where it grows.

ANGIOSPERMS

The largest division of vascular plants has seeds that are protected in some sort of a fruit (it may or may not be edible though). Included are woody shrubs, bushes, vegetables, herbs, grasses and many other flowering plants. This group also includes broadleaf trees, some of which are **deciduous**, which means they drop their leaves in autumn.

Identification of angiosperms includes characteristics of the leaves, fruits and seeds, but most of all, the flowers. Flowers may or may not have:

1 - outer leaflike **sepals** to protect the flower bud
2 - **petals** which may be colorful to attract insects
3 - female flower parts, together called the **pistils**
4 - male flower parts, together called the **stamens**

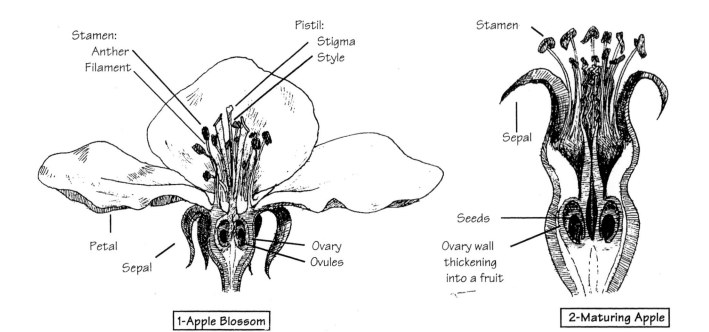

Vascular Plants

PARTS OF PLANTS

The six main parts of vascular plants are:

1 - roots   4 - flowers
2 - stems   5 - fruits
3 - leaves   6 - seeds

Fibrous root system

Roots absorb water and minerals from the soil and hold the plant in place.

Stems support the upper parts of the plant and transport food, water and other materials throughout the plant.

Tap root system

Leaves make food by the process of **photosynthesis**. The veins in leaves are vascular tissue which supply the leaf cells with water and minerals and carry the food away to the rest of the plant.

Flowers are the reproductive part of plants. Most plants have both male and female parts on the same flower, though some have them on separate flowers. The **pistil** (the female part) usually includes: **ovules** or eggs, protected in the ovary, the womb of the flower. It is usually at the base of a long tube called the **style**, while at the top of this tube is a sticky, pollen-collecting tip called the **stigma**.

The male parts of the flower are the **stamens** which produce dust-like **pollen** from the **anthers**, which are at the ends of long, thin supporting **filaments**.

Virginia creeper

REPRODUCTION

Three stages of reproduction are:

1 - pollination
2 - fertilization
3 - germination

When pollen gets to the tip of the pistil (an action called **pollination**), and the sperm from the pollen unites with the egg (an action called **fertilization**), the egg matures into a **seed**. The ovary around the seed or seeds enlarges and ripens into a **fruit**. The fruit protects the seeds, and may help spread them by wind, water or animals. A seed has: **1)** a protective outer coat; **2)** a tiny, living baby plant inside, called an **embryo** and **3)** food to get the baby plant started after it sprouts (an action called **germination**). Soon the seedling can grow leaves to make its own food, and roots to absorb water and nutrients.

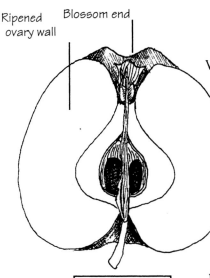

Ripened ovary wall · Blossom end

3- Mature Fruit

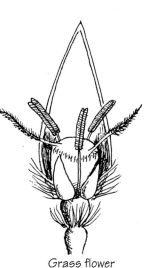

Grass flower

PHOTOSYNTHESIS

Castor bean leaf

Plants are different from most other living things because they can make their own food instead of having to eat. This process is quite complex, but basically it can be described as follows: carbon dioxide (**CO_2**) from air enters through the leaves, and water (**H_2O**) enters mainly through the roots. The chlorophyll in the cells uses energy from light to take these chemicals apart and put them back together as a different chemical, **glucose** (sugar) which is a simple food. Oxygen (**O_2**) is given off as a by-product of this process.

Western hemlock

$$CO_2 + H_2O + Energy \xrightarrow{chlorophyll} Glucose + O_2$$

Photosynthesis: Simplified chemical reaction

RESPIRATION

Poplar

Respiration is a basic process of all living things. It is also somewhat like the reverse of photosynthesis. Oxygen and food (glucose) are taken into the cells, where the sugar is taken apart to release energy needed for moving, growing and living in general. The by-products of respiration are carbon dioxide and water.

$$Glucose + O_2 \longrightarrow CO_2 + H_2O + Energy$$

Respiration: Simplified chemical reaction

PLANT GROWTH

Young bean plant

Plants have six basic needs for growing; they are:
- **1** - water
- **2** - light
- **3** - air
- **4** - proper temperature
- **5** - minerals or elements
- **6** - space to grow

Phototropism: Stem grows toward light

Geotropism: Roots are drawn downward

Water is needed for support, metabolic activities, transport of nutrients and photosynthesis. Light and air are needed for photosynthesis. Minerals or elements are needed for healthy growth and development of all plant tissues. Nitrogen, phosphorus and potassium are especially important, and are the most common elements in fertilizer.

When a seed germinates, the roots grow down and the stems grow up. This is because plants have chemicals in them called **hormones** which cause the plant to respond to its environment. Stimuli are what plants respond to such as gravity, light and water. **Tropisms** are the actions of growing toward or away from these stimuli. For example, when plant stems grow, there are hormones that cause them to grow toward the light. The stimulus is the light, which is what **photo** means, and so this growing toward light is called **phototropism**. When the stimulus is the pull of gravity from the earth, which is what **geo** means, then the growth of roots downward is **geotropism**.

Oak

Vascular Plants

EXAMPLES OF COMMON ANGIOSPERMS (FLOWERING PLANTS)

Umbelliferae: Queen Anne's Lace

Compositae: Sunflower

Family	Scientific Name	Examples	Characteristics
Grass	Graminae	Wheat, Corn, Oats, Rye, Barley, Rice, Bamboo, Millet, Sugarcane	Small flowers known as florets Usually grouped in spikelets May be arranged in spikes or heads The fruit is the typical grain
Pea	Leguminosae	Pea, Soybean, Bean, Peanut, Clover, Lupine, Wisteria, Locust, Alfalfa, Lentils, Vetch, Chickpea	Seeds in pods Butterfly-like flowers: 5 petals joined An upper banner 2 wings at sides 2 wings below 10 stamens
Parsley	Umbelliferae	Carrot, Dill, Parsley, Celery	Flower head like an umbrella 5 petals alternating with 5 stamens
Rose	Rosaceae	Rose, Almond, Peach, Pear, Apple, Cherry	5 petals and 5 sepals 10 to many stamens 1 to many pistils
Sunflower	Compositae	Sunflower, Thistle, Lettuce, Artichoke, Dandelion	Flowers grouped in heads: Outer ray flowers look like petals Smaller disc flowers in the middle One flower head is made up of many individual flowers

Ray flower

Disc flower

PROTOZOA
(to the tune of "Listen to the Mockingbird")

Septimus Winner — Lyrics by Doug Eldon

Amoebas are a sarcodine
Pseudopods to move are seen
False feet arrange the shape to change
And remember that the sarcodines are called:

All around the paramecia
You can see some moving cilia
Tiny hairs are wiggling there
And remember that the ciliates are called:

At the tail end of the Euglena
They will all have a flagella
The tail to whip is from the tip
And I know that the flagellates are called:

Sporozoans live in a host
Movement's limited the most
Malaria's carried by a mosquito
The plasmodium is moved to a new host

PROTOZOA

Protozoa also called protista are microscopic organisms that were considered animals until fairly recently. The only real difference is that protozoa are only one cell, whereas animals are many cells all working together.

Protozoa usually live in water, but many live inside animals. They feed by absorbing, ingesting or even by making their own food. There are several ways protists reproduce, but the most common is by dividing in half (**binary fission**).

Most protozoa have some way of moving. In fact, they have been grouped according to the way they move.

SARCODINES

The *amoeba* is the best known sarcodine. It has even inspired science fiction writers and filmmakers, with such classic movies as "The Blob."

Amoebas do not have a particular shape, but it changes, which is how they move and how they catch their food. Part of their blob-like bodies extends forward, forming what is called a "false foot" or **pseudopod**, which is the characteristic of this group. To move, these extensions pull the rest of the body along. To catch food, such as bacteria or other protists, the pseudopod first surrounds the prey, which is then digested in a structure called a **vacuole**. What is usable is absorbed through the cell membrane and what is not usable is excreted as waste.

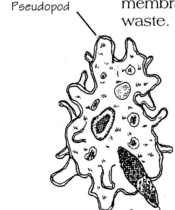

Although most amoebas live in fresh water, some live in animals. In fact, three out of four people are thought to have harmless mouth amoebas. But in warm or tropical climates there is a harmful amoeba that enters people's digestive systems, causing amoebic dysentery. This amoeba is spread in the form of a spore-like cyst in human excrement, which can then contaminate food, drink or objects that may be put in people's mouths. The symptoms are fever, severe and bloody diarrhea, along with

An amoeba preying on a flagellate

swelling and pain. Infection can be prevented by purifying water, washing hands and food thoroughly, and by general cleanliness and sanitation.

CILIATES

Ciliates are protists that have tiny hairlike **cilia** that move back and forth. On some, such as the slipper-shaped **paramecia**, the cilia are all over the body and propel the little creature around. The cilia also move food particles toward the **oral groove**. On other ciliates the body is shaped like a trumpet or funnel and the cilia create a whirlpool around the opening, sucking food in.

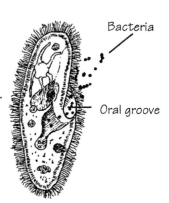
Paramecium

Stentor

FLAGELLATES

The characteristic of the flagellates is a tail-like **flagellum** that whips back and forth making it possible for some flagellates to swim rapidly. **Volvox** is a colony of many individuals, each with its flagellum facing out and whipping together, giving the ball-shaped colony the ability to roll around in its watery environment.

There are some flagellates that are able to make their own food, much as plants do, while others are parasitic. Other kinds live in the stomachs of termites, digesting the wood the termites eat. The insects cannot live without this protozoa in their digestive system.

Volvox

Euglena is a flagellate that is able to make its own food. It is unique because in addition to having chlorophyll, it can also ingest (eat). This common protist has an oval shape, pointed at one end and rounded at the end with the flagellum. Euglena often forms a layer of green scum on the surface of ponds.

Euglena

SPOROZOANS

Sporozoans are parasitic protists that are not able to move by themselves. With no pseudopods, cilia or flagella, these simple protists are carried by their **hosts** (the animals they live in) to different places. They are named sporozoans because they reproduce by making spores.

The most important member of this group is also considered the most dangerous living thing in the world! The organism is named **Plasmodium** and the disease is **malaria**. More people have had this disease than any other; it is the leading cause of death!

Not counting wars or accidents, this tiny sporozoan is probably responsible for half of all deaths since prehistoric times![8] Even in modern times 200 million people a year get malaria, mostly in tropical climates. The spores are carried from an infected person by a particular species of mosquito. When that mosquito bites another person, plasmodium is then passed on and the disease is spread.

The plasmodium eventually grows and multiplies in the person's red blood cells. Symptoms are fever, chills and sweats, great weakness and eventual death. There are treatments for those with malaria but the best way to control the spread of the disease is to control the mosquito that carries the plasmodium.

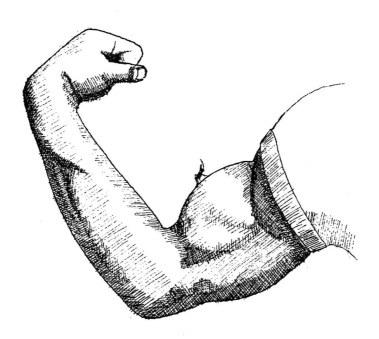

Transfer of plasmodium

GENETICS
(to the tune of "Shortnin' Bread")

DNA is what they say, made you what you are today. Like a program it will tell what will happen in each cell. DNA is a molecule.... look real close it's really cool. Like a ladder in a double helix, so simple, yet so complex.

Chorus:
Deoxyribonucleic acid, Talkin' 'bout genes and chromosomes. Deoxyribonucleic acid Talkin' 'bout genes and chromosomes.

In chromosomes made of DNA
   There is a particular way
Nitrogen bases can be seen
   In special orders to make a gene

When organisms reproduce
   Different genes are on the loose
Dominant and/or recessive
   The offspring get traits the parents give

Genetics is the study of heredity,
   It may use probability
To predict just how likely
   Results of genetic crosses will be

Genetic engineers can use
   Parts of different molecules
Changing DNA around
   To solve a problem they have found

GENETICS

Genetics is the study of **heredity**—how offspring **inherit** characteristics (**traits**) from their parents.

An Austrian monk named Gregor Mendel is considered the father of genetics for the work he did with peas in the 1860's. He experimented with peas that had different traits: seed shape and color, pod shape and color, stem length, etc.

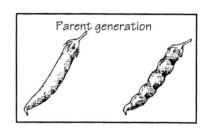

He found that if he pollinated flowers of a plant with one trait, such as full pods, with pollen from flowers of a plant with another trait, such as pinched pods, the first plants that grew all had full pods. But when he did the same with the flowers of those pea plants, he found that most had full pods, but some had pinched pods.

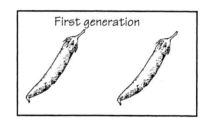

The same thing held true with the other traits: the first generation resembled the parent generation while the second generation was more diverse. After many experiments and a lot of thinking, Mendel concluded that each trait was determined by two factors. The factors that determine the traits the offspring get from their parents are now called **genes**. They come in pairs, and each parent contributes half of each pair.

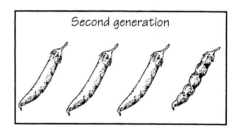

In a pair of genes, there is often one trait that shows up more than another, and is therefore considered **dominant**. The gene that determines a trait that does not show up as often is considered **recessive**.

Traits from recessive genes are hidden or masked by the dominant genes. An example of this is in a family where one parent has dark skin, dark eyes and dark hair, and the other parent has light skin, blue eyes and blond hair. The dark genes are dominant, so most of the children in this family will have the darker traits. However, some of the lighter traits will likely show up in one or more of the children, or else in the grandchildren.

Genes were discovered to be located on the thread-shaped structures found in the nucleus of cells, called **chromosomes**.

Chromosomes

In 1953, two scientists were able to describe the chromosomes as being a complex molecule, **deoxyribonucleic acid,** or **DNA**. This molecule is made up of thousands of atoms and, if described in simple terms, has the shape of a ladder that has been twisted. The twisted ladder shape is called a **double helix**. (A **helix** is something twisted into a spiral; because DNA has two strands, it is a double helix.)

DNA is in many ways like a computer program. When a disk is loaded into a computer, it instructs and controls what the computer does. In the same way, DNA instructs and controls what goes on in the cell. It controls the production of proteins which is what determines the traits or characteristics.

You look the way you do because of the "code of life"— the DNA in your cells! An individual's DNA is unique; people can now be identified by their DNA, instead of by their fingerprints, which has become useful in solving crimes.

The "rungs" on this twisted ladder are each made of a pair of substances called **nitrogen bases**. A base is a kind of chemical with particular properties, and nitrogen is a kind of atom that is part of these bases. There are only four different kinds of nitrogen bases in DNA, but there are millions of pairs on each DNA molecule. The order of the pairs make up a gene; the particular order is what determines the traits the offspring get from their parents. It is so simple, yet so complex!

Probability is a mathematical term used to describe or predict how likely or how possible it is for an event to take place. In genetics, especially in plant and animal breeding, this is a way to figure out ahead of time what the chances are of certain traits showing up in offspring.

GENETIC ENGINEERING

Genetic engineering involves the transfer of genes, or parts of DNA molecules, from one organism to another. So far, this has been from complex organisms such as humans to simpler organisms, especially yeast and bacteria. For example, a section of human DNA can be spliced into a DNA molecule of a bacteria. The combination of the two is then called **recombinant DNA**. The bacteria with this mix will then be producing what the other organism normally would, except much faster.

By combining parts of different DNA molecules genetic engineers have done some incredible things: producing vaccines; human insulin (for treating people with diabetes); plants that are resistant to diseases; a kind of harmless bacteria which can be sprayed on plants to protect them from freezing; and some possible treatments for cancer, AIDS and other diseases. The work of changing DNA around to solve problems and make improvements in living things is called **biotechnology**.

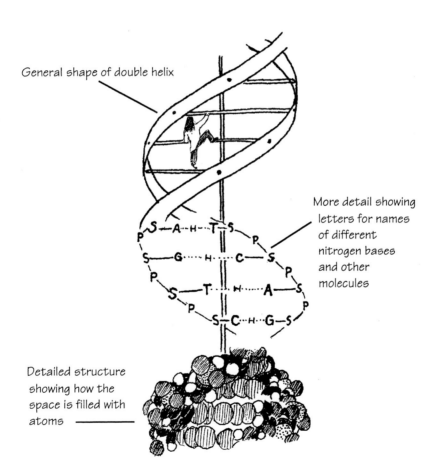

VIRUSES
(to the tune of "Yankee Doodle")

Richard Schuckberg

Lyrics by Doug Eldon

Vi-rus-es cause man-y dif-ferent in-fec-tious dis-eas-es
In-flu-en-za com-mon colds (with fe-vers, coughs and snee-zes).
Ra-bies, vi-ral pneu-mo-nia, and mo-no-nu-cle-o-sis
Chick-en pox, mea-sles and mumps, and AIDS, con-junc-ti-vi-tis.

Chorus
Vi-rus-es are ver-y small, ex-treme-ly mi-cro-sco-pic.
Few of them are good at all from them the host may be-come sick.

Viruses are not living, but are just tiny particles
Can't perform life functions; are not cells and don't contain cells
Reproduce only within a living cell and will take most
Of that cell's own processes to multiply inside a host

In the core of viruses you'll find nucleic acids
Molecules which do control production of new viruses
RNA or DNA, under close inspection
Are surrounded by a coat of protein for protection

VIRUSES

INFECTIOUS DISEASES

An infectious disease is a sickness passed from one organism to another; it invades the organism, causing harm in some way. Another name is **communicable disease**, which means **contagious** or easily spread.

Most infectious diseases are caused by harmful microorganisms (microbes) that are referred to as **germs**. They are mostly viruses or bacteria, and can be **transmitted** (or spread) most often by:

   1 - direct contact with the infected person.

   2 - droplets in the air (especially from the infected person sneezing or coughing).

   3 - eating food or drinking what has been contaminated. That is why hand washing and cooking food completely are so important to the control of infectious diseases.

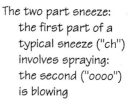

The two part sneeze: the first part of a typical sneeze ("ch") involves spraying: the second ("oooo") is blowing

CHARACTERISTICS OF VIRUSES

A virus is not really a living organism! It is not able to perform life functions:
- it is not able to reproduce on its own
- it does not move
- it does not grow
- it does not respond to stimuli
- it does not carry on metabolic activities

A virus is not a cell and does not contain cells. Instead, it is just a tiny particle that can be a hundred times smaller than most living cells!

Although the size and shape may vary, a virus is amazingly simple, having basically two parts:

1 - A core of genetic material made up of molecules that control reproduction, known as **nucleic acids**. Nucleic acids are particular kinds of chemicals found in the nucleus of cells and include ribonucleic acid (RNA) and deoxyribonucleic acid (DNA).

2 - A protective coat of protein surrounds the core.

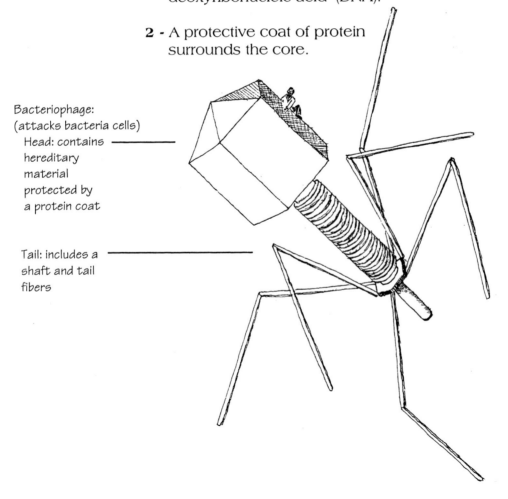

Bacteriophage: (attacks bacteria cells)
Head: contains hereditary material protected by a protein coat

Tail: includes a shaft and tail fibers

VIRAL DISEASES

Diseases caused by viruses are some of the most contagious, hard to control, difficult to treat and deadly.

Influenza, or **flu** for short, has been one of the most serious viral diseases. In 1918, an **epidemic** (or rapidly spreading disease) of influenza became a **pandemic** (an epidemic that infects many people over a wide area). In 16 months, over 20 million people died! Even now, each year in the United States influenza causes 20,000 to 40,000 deaths.

A fever is one of our defenses: germs do not live and reproduce as well when our body temperature is high.

The **common cold**, though not as serious as the flu, is by far the most widely spread viral disease.[9] There are more than a hundred different kinds of viruses that affect our noses and throats, and most of us catch two or three colds a year.

Rabies is caused by a virus that infects wild animals and dogs and then can infect humans. It affects the brain in such a way that the rabid animal goes crazy, or mad, and will be more likely to bite. In this way the virus in the animal's saliva is transmitted to a new animal.

The name rabies comes from the Latin word meaning rage or madness; it has also been called **hydrophobia.** This word means fear of water because a person with the disease has trouble swallowing. Rabies was one of the most feared diseases before Louis Pasteur developed a vaccine for rabies.

Pneumonia can be caused by either a virus or a bacteria, but both are lung diseases. A big difference is that viral pneumonia cannot be treated and bacterial pneumonia can be treated with antibiotics.

Infectious mononucleosis, or **mono**, affects the blood, causing an increase in the monocytes of the white blood cells and raising the number of abnormal white blood cells. The symptoms include fever, sore throat, loss of energy and enlarged lymph glands, especially in the neck. Mono is less severe than many other infectious diseases. It most often affects young adults, and is commonly transmitted by direct contact.

Chicken pox is a common childhood disease, as are **mumps** and **measles**. **German measles**, (or **rubella**) is another highly contagious viral disease that children

Measles

do not often get any more because they get vaccines that make them immune (see chapter on bacteria). In fact, before entering school, children in the U.S. must have had **MMR** vaccines (for Measles, Mumps and Rubella). In this way, fewer children get these diseases, and the virus is less likely to be spread. A vaccine for chicken pox has now been developed which will also control this childhood disease.

AIDS stands for **Acquired Immune Deficiency Syndrome**. **Acquired** means something you get, or catch from someone else; **immune** refers to our body's system of fighting off diseases with different kinds of cells, such as white blood cells, that attack and destroy invading viruses or bacteria; **deficiency** means a shortage or something important missing; and a **syndrome** is a number of symptoms, or signs of a problem, that occur together.

Conjunctivitis

The virus that causes AIDS is the **Human Immunodeficiency Virus**, or **HIV**. This virus attacks the cells that defend our bodies from other diseases, making the immune system unable to fight off bacteria that do not normally cause problems for a healthy person. Someone with AIDS ends up getting very sick, not so much from the HIV as from pneumonias and certain cancers that people with effective immune systems usually do not get.

Influenza

Conjunctivitis is **inflammation** (or swelling) of the sensitive tissues around the eyes, caused by a virus or a bacteria; it is commonly called pinkeye.

Other viral diseases include:

Poliomyelitis (or **polio**) crippled many people, especially children, until a vaccine was developed in 1954. Fewer people have been affected by the polio virus since then.

Polio

Smallpox used to be a very serious disease. It was brought to the Americas by European explorers and early settlers. Native Americans had little immunity to it; whole tribes were wiped out from this virus! With the development of a vaccine and world-wide efforts, this disease has been eradicated.

Yellow Fever

Yellow fever is caused by a virus that is carried by a particular kind of tropical mosquito. The disease has been controlled by killing the mosquitoes that transmit the virus, and by developing a vaccine that helps our bodies fight off the virus.

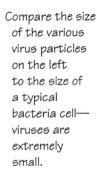

Compare the size of the various virus particles on the left to the size of a typical bacteria cell— viruses are extremely small.

Bacteria

Foot and mouth (or hoof and mouth) **disease** is extremely contagious, especially among cattle, sheep, hogs, and deer. The main symptom is blisters on the feet and in the mouth. The virus rarely attacks man.

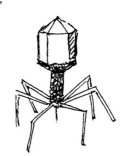

Fifth disease is so named because it was the fifth disease to be found to cause a skin rash, especially among children.

Hepatitis is a name given to several kinds of diseases, some are quite serious and most affect the liver. There are vaccines for some types of hepatitis.

Other viruses attack plants, causing disease and lower productivity. **Tobacco Mosaic Virus (TMV)** infects tobacco and related plants, including tomato and potato.

REPRODUCTION

When a virus infects an organism, it enters a particular target cell. To get inside the host cell the nucleic acids leave the protein coat and travel into the cell to the nucleus. Then the nucleic acids (the viral RNA or DNA) take over the host cell's production of proteins. The host cell now makes what is needed for the virus's protective coats, and the cell's reproduction processes are commanded to make more viral RNA or DNA! The virus has the living cell reproduce viruses instead of more living cells!

At some point, the living cell dies, but by that time it may have produced several hundred to several thousand new virus particles which burst forth to infect other nearby cells.

Part of the problem in trying to treat viral diseases, or at least control them, comes from the fact that they are not really living. Antibiotics are good at killing living things such as bacteria, but they do not affect viruses. Vaccines have been developed for many viruses but some, particularly HIV, are able to change so fast that the immune system (even with the help of vaccines) just cannot keep up. For being so simple and so very small, viruses are certainly a serious threat for living things.

Conjunctivitis (Large scale)

OH BACTERIA
(to the tune of "Oh Susanna")

Stephen Foster   Lyrics by Doug Eldon

For decomposing things that die, a saprophyte we need
But some are parasitic, on a living host will feed
For taking nitrogen from air, and fixing it into
The soil for plants to use, I'm sure they're all grateful to you

Chorus:
Oh bacteria though simple and so small
Without you ecosystems would not function well at all

In dairy products you have shown yourself a cultured friend
And to genetic engineers, your DNA you lend
You even help to fight diseases caused by your brethren
You make antibiotics which destroy or weaken them

Chorus:
Oh bacteria, though only single-celled
A most important organism we have now beheld

Though most of you are helpful, in some of these mentioned ways
There are a few who have to do, a bit with some disease
Producing toxin or the cells attacking directly
Diphtheria, pneumonia, strep throat, tetanus, and TB

Chorus:
Oh bacteria, though only single-celled
A most important organism we have now beheld

We do appreciate you and your praises we do sing
Yet some of you make life so hard with troubles that you bring
Our food you spoil, our crops you rot, our animals attack
With botulism, different rots, cholera and anthrax

Chorus:
Oh bacteria, though only single-celled
A most important organism we have now beheld

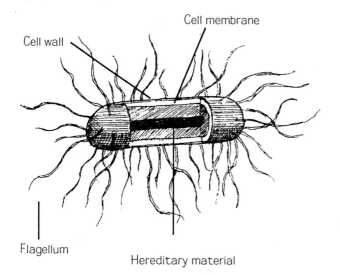

BACTERIA

Bacteria are living, single-celled organisms. Each cell lives independently, and is able to carry on metabolic activities, respond to stimuli, and reproduce. Bacteria have cell walls, free floating genetic material but no nuclei; this makes them different from cells in the other kingdoms.

Bacteria can live almost anywhere. Often they survive where no other living thing can: in the upper atmosphere; six miles deep in the ocean; in frozen soil; in hot springs; and even where there is no oxygen at all. There are different kinds of bacteria living in and outside of us. In fact, it is bacteria that are responsible for body odor, bad breath, cavities and many infections.

All bacteria need water, and most need food and oxygen. But when conditions are not good, and these needs are not being met, many bacteria produce a hard outer structure called an **endospore**. This protective structure has been found to be resistant to almost anything! When conditions improve, the endospore opens and the bacteria again become active.

REPRODUCTION

Reproduction in most bacteria is simply by dividing in two, called **binary fission**. This can happen as fast as every twenty minutes. That means that if all survived, after seven hours there would be over a million! Fortunately, not all survive but the speed of reproduction is important, both with helpful and harmful bacteria. **Decomposition** of such things as sewage is rapid because bacteria multiply rapidly; but for the same reasons, infections and diseases can spread at alarming rates.

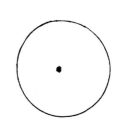
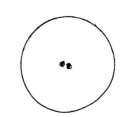
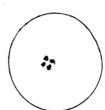
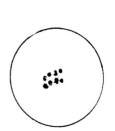
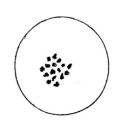

Binary fission is multiplication by division! (The more it divides, the more there are)

Bacilli

Bacteria are very small, much smaller than any other living thing. There are millions in a pinch of soil, in a drop of water, even in your mouth! They couldn't be seen until microscopes were developed, and many scientists laughed at the theory that there were tiny germs that were responsible for diseases. But by the late 1800's, a whole new field of science was developing: **microbiology**, the study of microscopic life. Later, as microscopes improved, **bacteriology** became a separate field itself.

CLASSIFICATION

Bacteriologists classify bacteria by shape into the following three main groups:

1- **Bacilli** are rod-shaped, often with whip-like flagella.
2- **Spirilla** are spiral or comma-shaped.
3- **Cocci** are spherical. If the cocci are linked in chains they are in the subgroup **streptococci**, and if they are in bunches like grapes they are **staphylococci**.

Spirilla

IMPORTANCE

An ecosystem is made up of all living and nonliving things working together. Nonliving things include air, water and minerals in the soil. Living things are described, as discussed earlier, as either producers, consumers or decomposers.

Producers are plants, which make food using energy from the sun.

Consumers eat either the plants or other consumers.

Cocci

Decomposers break down dead plants and animals, returning minerals back to the soil. Decomposers complete the cycle that was begun when plants absorbed nutrients from the soil. An organism that gets energy from decomposing dead matter is called a **saprophyte**. Bacteria and fungi are the most important; without them we would quickly be buried in tons of dead plants and animals!

A **parasite** gets its energy by feeding on a living organism called a **host**. We act as hosts to many bacteria, some of which end up causing us harm by producing infections and diseases.

HELPFUL BACTERIA

A group of bacteria known as **nitrogen-fixing bacteria** are important to plants. Nitrogen is a substance needed by plants and often lacking in soils. Air would be a good source of nitrogen—it makes up 78 percent of air—except that it is in a form that the plants cannot use. The helpful nitrogen-fixing bacteria live near the roots of certain plants, especially legumes (peas, beans, clovers and lupines), forming little bumps called **nodules**. They take the nitrogen from the air and change it, "fixing" it into the soil in a way that it becomes usable to plants, much like a fertilizer.

Bacteria are important in certain foods we eat, especially dairy products. By adding a **culture** of a particular bacteria to milk we get cheeses, yogurt, sour cream, or buttermilk.

There are also bacteria that normally live in the human intestine which actually help digestion. If antibiotics are taken as a medicine to kill harmful bacteria, the helpful bacteria are killed too. Eating certain foods that contain bacteria like yogurt build up the intestinal bacteria population again. There is a species of harmless bacteria that lives in human intestines called *Escherichia coli*. There are specific strains, however, which if ingested will cause food poisoning—diarrhea, vomiting, possible kidney failure and strokes. Such is the case with the dangerous strain of *E. coli* that people have contracted from eating contaminated food.

Cattle, sheep and goats need certain bacteria in their digestive systems in order to break down grass and make it digestible.

Bacteria have also been found to be helpful in disposing of sewage, digesting oil spills, producing biodegradable plastics, and controlling pests. One pest-controlling bacteria is *Bacillus thuringiensis*, or B.t. which kills caterpillars but is harmless to all other animals. A certain stain of B.t. has recently been found which attacks malaria-bearing mosquitoes.

As discussed previously, another helpful use of bacteria that is fairly new is called **genetic engineering**. Scientists found that they could take DNA from bacteria, cut it up and splice it together again in a way that the bacteria would then produce something useful to man. An example includes the production of insulin, needed by people with diabetes.

BACTERIOLOGY: A Short History

Although many bacteria have been found to be beneficial and even essential, other kinds are now known to cause problems. Yet the knowledge and understanding of bacteria, especially of those that cause diseases, has come within the last 150 years.

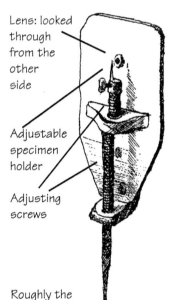

The first microscope

Lens: looked through from the other side

Adjustable specimen holder

Adjusting screws

Roughly the actual size of Leeuwenhoek's hand-held microscope

Father of Microbiology

Anton Leeuwenhoek
Tiny organisms were first seen about the year 1700 by a man from the Netherlands named Anton Leeuwenhoek (pronounced LAY-vun-hook). He called these microscopic forms of life **animalcules**, and tried to convince scientists that many things are affected by them. He is considered the Father of Microbiology because he invented the first microscope. The connection between germs and diseases, however, did not come until the later part of the 1800's. Several scientists, using scientific methods and microscopes, revealed a whole new world of living things. This was to change science, as well as health and medicine, forever.

Fathers of Bacteriology

Robert Koch (pronounced Cockh), a German doctor, and Louis Pasteur, a French chemist, are each called the "Father of Bacteriology," and each made incredible contributions worthy of discussion.

Robert Koch
Anthrax, the deadly disease of cattle and sheep, was discovered by Koch to be caused by a specific bacillus. He found that once in the blood of an animal, this thread-like bacillus would quickly multiply, killing the animal in about one day! When the animal was dead the anthrax bacteria would form endospores that would infect other animals. Entire herds and flocks were being wiped out in Europe by anthrax.

Cholera is a very infectious intestinal disease that is still killing thousands of people. It was also discovered by Koch to be caused by a particular bacteria. Epidemics of cholera had come and gone for centuries; all that was known about it was that once a person had it, they would die in a day or two.

In the early 1800's a scientist in London was able to trace the source of an epidemic to a particular water pump. It then became clear that contaminated water spread this disease, but Koch found the real cause of this dreaded disease: the cholera bacteria in the water. He was able to show that it was also spread by filthy conditions and poor sanitation.

Tuberculosis is a disease that Koch discovered was caused by the tubercle bacillus. The discovery of **TB** made Koch famous. For hundreds of years this was the most dreaded disease; no one knew what caused it or how it was spread, but for centuries it claimed the lives of three out of every seven people in Europe. TB is also called "**consumption**" because people who contracted this disease would gradually waste away and finally die. Koch identified the responsible microbe and the fact that it was spread by breathing in the bacteria, as when an infected person sneezed or coughed.

Koch developed scientific methods of finding which organisms caused diseases and how to treat them. His rules of studying the causes of diseases are now known as Koch's postulates. He also trained other scientists who later found causes and treatments for many serious diseases, including bubonic plague, diphtheria and tetanus.

Louis Pasteur
Louis Pasteur was a French chemist who became involved with wine and beer makers who were having trouble with their products fermenting properly. Instead of producing alcohol and normal flavors, the brewers were getting vinegar and souring. Pasteur found that normal fermentation was from a certain yeast (a kind of fungi), and the souring was from a bacteria. He then developed a way to kill the bacteria: he simply heated the brew to around 150 degrees for a few minutes. We now call this process of heating to kill bacteria **pasteurization**. Pasteurized milk is safer to drink than raw milk because harmful bacteria are destroyed. Apple juice that has been pasteurized won't ferment. Apple cider is different in that it has bacteria as well as yeasts in it, which will cause it to ferment, producing vinegar or alcohol.

Vaccines

Pasteur is also famous for developing a way to keep animals from getting cholera and anthrax. He found animals would develop a natural way of fighting diseases if given shots of weakened disease organisms. The shots, called **inoculations** triggered the production of **antibodies**. These tiny particles would be produced to attack specific germs and would make the animal immune to that disease. When the animal was infected or exposed to those germs, it would have the defenses to fight them off without getting sick.

The shots are called **immunizations** because they make the animal immune to that disease. They are also called **vaccines**. Because of Pasteur's work, children are vaccinated at an early age so they will not get many diseases people used to get.

Germ Theory

Pasteur developed the theory that microscopic germs were all around us, especially on the dust in the air, and that these microbes could cause disease, decay, and fermentation. He also suggested that doctors and surgeons were actually spreading diseases by their practices. At that time, when surgery was done, gloves were not worn and hands were not washed. Wounds were not cleansed and the same instruments (tools) and bandages were used on different patients. Hospitals were filthy places where people died more often from infections than anything else. But because Pasteur was not a doctor, and because this "germ theory" was new, he was neither listened to nor believed.

Father of Antiseptic Medicine

Joseph Lister

It was not until the late 1860's that a Scottish doctor, Joseph Lister, took the germ theory seriously. He began using a kind of acid that killed germs on everything: the instruments, tools, and equipment; hands of all doctors and helpers; on the wound itself; and on the bandages and dressings. This was the beginning of **antiseptic** medicine— sterilizing, or killing germs, or at least keeping them from causing harm. By keeping his hospitals clean and germ-free, Lister reduced the death rate from fifty out of every hundred to about three! He came to understand how well our bodies can heal themselves, if infections are prevented. By applying what Koch and Pasteur had discovered, Lister saved countless lives.

Because of him, we now have many products that kill or weaken harmful bacteria and prevent infections:

- sterile bandages
- disinfectants
- antibiotics
- antiseptic ointments
- antibacterial soaps
- mouthwashes

Vaccines and immunizations further help our bodies fight infectious diseases caused by bacteria or viruses. Even controlling flies and other germ carriers help control the spread of diseases. We have Robert Koch, Louis Pasteur and Joseph Lister to thank for much of our present knowledge about bacteria, other microorganisms, and diseases. Medicine would not be what it is today without their work.

BACTERIAL DISEASES

In addition to anthrax, cholera, tuberculosis, and infections, there are other bacterial diseases which have been, or still are very serious:

Bubonic Plague and two other forms of plague caused the worst disaster in history.[10] In four years, from 1347 to 1351 one fourth of Europe's population died from what became known as the Black Death! The bacteria was carried by fleas that lived on rats which lived in the filthy city streets. At the time, people didn't have a clue as to what caused the disease, or how it was spread; but anyone who got the plague was sure to die. We now have antibiotics and drugs for treatments, and we know how to control the source—the rats and the germ-carrying fleas.

Diphtheria most often affects infants and is caused by bacteria that live in the nose and throat being spread by coughing or sneezing. There were epidemics of diphtheria in the late 1800's, but there have not been many cases since then. Recently, however, epidemics have been spreading in Russia where few people have ever been inoculated with a vaccine.

Tetanus is also called **lockjaw** because it causes the jaw muscles to become paralyzed. The bacteria live in the soil and can enter the body through cuts or a puncture wound (such as stepping on a dirty nail). Most of us have had a tetanus shot to make us immune, and get a booster shot every ten years or so to keep our resistance strong.

Pertussis, or **whooping cough** is a highly contagious disease that can be very serious, particularly for children. As the name suggests, the symptoms are a "whoop" cough (that sounds like the caw of a crow when breath is drawn in) and severe coughing spells that make eating, drinking, and breathing difficult. Pertussis lasts for one to three months, and often brings on pneumonia. A vaccine given to most children to make them immune to the bacteria is usually combined with vaccines for diphtheria and tetanus. This vaccine is called **DPT**, for Diphtheria, Pertussis, and Tetanus; and is given three times in the infant's first six months to build up immunity.

Pneumonia is a disease affecting the lungs, sometimes following a bad cold, the flu, or some other respiratory disease. Bacterial pneumonia can be treated and is not usually fatal. Viral pneumonia, however, cannot be treated with antibiotics.

Strep throat is a common contagious sickness causing a severe sore throat, but can be treated with antibiotics.

Dysentery is a disease that occurs in all climates and all countries, especially during the summer. It is spread by contaminated food, and is found where large groups of people are living and eating together and where sanitation is poor. The symptoms are severe diarrhea, stomach cramps, and fever. **Typhus** is similar in many ways. Both diseases can be kept from spreading by washing hands, food and cookware; by cooking food well; and by proper disposal of human waste.

Botulism is another bacteria that lives in the soil, but this one can cause a deadly food poisoning. If food is not preserved properly, the bacteria can multiply in the can or jar, producing a **toxin** (poison) which when ingested by people gets into the blood. Gas is also produced, which makes the top of the can (or lid of a jar) bulge—a good warning! Treatments are available, but heating and sealing containers completely is usually all that is necessary to prevent botulism.

There are many other bacterial diseases that attack plants causing them to get sick and die. Some of these can spread rapidly, such as **fire blight**, which has killed whole orchards within a few weeks.

Rots are usually related to fungi, but those caused by bacteria can be especially slimy and foul-smelling. Bacterial rot of potatoes is a good example.

Appendix

Dodo

THE SCIENTIFIC METHOD
(to the tune of "Dixie")
Lyrics by Doug Eldon ©

Oh, what do you think a scientist does
To solve a problem found because
Many scientists are scientists
'Cause they're great problem solvers

There is a systematic way
They go about 'most every day
It's methodical and it's logical
The scientific method

Chorus: A way to solve a problem, a way, a way
The scientific method is a way to solve a problem
A way, a way, a way to solve a problem
A way, a way, a way to solve a problem

It may not seem important to you
But the first thing that they always do
Is state the problem or ask a question
So they know just what they're after

Then they review everything involved
That might help get the problem solved
By reading, researching
And gathering information

Chorus

After both of these steps they take
They go ahead and then they make
An educated guess—a hypothesis—
A possible solution

Then they use scientific tools
To measure and test some variables
In experiments which are really meant
To give more information

Chorus

This information they call data
They put together so that later
They can analyze and synthesize
To see just what it all means

Only when they have done all these
Experiments testing hypotheses
Which may prove, or else disprove
Then they'll state their conclusion

Chorus

This is the systematic way
A scientist may use any old day
'Cause it's methodical and it's logical
The scientific method

ALL LIVING THINGS
(to the tune of "I Love the Mountains")
Lyrics by Doug Eldon ©

All living things are able to reproduce
Move and grow and respond to a stimulus
And carry on metabolic activities
These are characteristics of living things

There are four metabolic activities:
Ingestion and digestion are two of these
Respiration and excretion
Metabolic, metabolic: chemical activities

Needs of living things include energy
Water, oxygen and food to eat
Living space and proper temperatures
All living things have these six basic needs

Living things are all made up of cells
Units of structure and functions you can tell
All cells come only from other living cells
This is what's called the cell theory

Cells that are similar joined together form tissues
Tissues working together form organs
Organ systems and organisms are:
Five levels in which living things are organized

Kingdom, phylum, class, order, family
Genus and species make the name you see
Nomenclature and taxonomy
Classify, classify, name and classify

Diatom: a type of golden algae

INVERTEBRATES

(to the tune of "Clementine")
Lyrics by Doug Eldon ©

Many creatures have different features
  Yet all have a common trait:
    with no backbone
    they are all known
    to be called invertebrate

All invertebrates together make a bit
  over 89 percent
    of the animal
    species in the world
    They live in any habitat

They're divided into phyla,
  by their structures classified
    Here are eight kinds
    if you do find
    can then be identified

Poriferans are really sponges
  and they all have tiny pores
    Cells in colonies
    they surely are pleased
    living on the ocean floor

Cnidarians like jellyfish, coral
  hydra, sea anemone
    cells for stinging
    have one opening
    hollow body cavity

Platyhelminths are the flatworms
  like the small planarians
    nematodes are
    little roundworms
    segmented worms are annelids

Echinoderms have spiny skins and
  tube feet coming out of them
    sea cucumber,
    starfish, sand dollar
    and the spiny sea urchin

Slugs and snails are one-shelled mollusks
  clams and scallops have two shells
    octopus, squid,
    nautilus are
    headfooted with tentacles

They all have a softish body
  with a mantle that can make
    a hard shell
    that you can tell will
    give protection for their sake

Arthropods have jointed legs
  and a hard exoskeleton
    Centipedes,
    millipedes, insects
    arachnids, crustaceans

Centipedes have fewer legs and
  they are also carnivores
    millipedes
    have many more legs
    scavengers and herbivores

Crabs and lobsters, shrimp and barnacles
  are crustaceans you can tell
    four antennae
    legs are many
    they're aquatic with a shell

Arachnids include the spiders
  Ticks and mites and scorpions
    They all have eight legs
    two body parts
    No antennae and no wings

Insects have six legs, three body parts
  two antennae and two eyes
    egg to larva
    change to pupa
    then to adult with wings to fly

Chambered nautilus

COLDBLOODED VERTEBRATES
(to the tune of "When Johnny Comes Marching Home")
Lyrics by Doug Eldon ©

Oh, when you study animals, there're some of which you're told
Whose blood will always stay quite warm;
But some whose blood is cold
They can survive within a range
Of temperature except a change
That's too extreme can be so dangerous for animals
For coldblooded vertebrates; coldblooded animals

They must respond by what they do and so they move around
To find the proper temperature
In water or on ground
Fish, reptiles, amphibians
Whose temperature of blood has been
Controlled from outside not within these kinds of animals
These coldblooded vertebrates; coldblooded animals

The fish have gills instead of lungs to get their oxygen
And most have air bladders and fins
To help them float and swim
Some are jawless, like lampreys
With cartilage like sharks and rays
But most have bony vertebrae, they're bony animals
They're coldblooded vertebrates; coldblooded animals

Amphibians lead double lives, that's how they get their name
They start in water, then go on land,
Which they like just the same
Toads have dry and bumpy skin
Frogs' is wet and smooth as in
The salamanders and newts, their kin. They're all amphibians
They're coldblooded vertebrates; coldblooded animals

Snakes and lizards are reptiles, along with tortoises
Whose legs are made for walking,
Not for swimming like turtles
Crocodiles have teeth that show
An alligator's will always go
Inside its mouth that's how to know these different animals
These coldblooded vertebrates; coldblooded animals

These animals are different in the way they reproduce
The fish must fertilize their eggs
Externally and loose
Amphibians' jelly-eggs are wet
Reptiles' are leathery and set
To hatch on land where they will get to become animals
These coldblooded animals; coldblooded animals

BIRDS

(to the tune of "If You're Happy and You Know It")
Lyrics by Doug Eldon ©

Animals that people study quite a bit
That are warmblooded and are also vertebrate
We do classify as birds
Now, we'll share with you some words
So some interesting facts you won't forget

Characteristics are a beak, two legs and wings
Several kinds of feathers which do different things
There are some that help them fly
Others make them more streamlined
Fuzzy, down feathers are insulating

Bones of birds are light because they are hollow
So it's easier to fly, and don't you know—
Both to cool them when they're flying
And for oxygen supplying
They have air sacs to help them on the go

Eggs let oxygen pass right on through the shell
Often laid inside a nest constructed well
The eggs the parents incubate
Hatch their young and may migrate
Following their food supply—there's more to tell

The class of birds has many orders and families
Which are divided into genus and species
But nine thousand different kinds
Are too many for your minds
So forget about more verses, if you please

Heron

ALGAE, FUNGI AND NONVASCULAR PLANTS
(to the tune of "Go Tell Aunt Rhodie")
Lyrics by Doug Eldon ©

Algae and fungi, lichen, moss and liverworts
All are nonvascular and reproduce by spores
Algae is classified by color into five groups
They can be green, or brown, golden, red, or fire

Fungi lack chlorophyll, they get energy other ways
Most by decomposing, or fermentation
If they live on dead things, they are known as saprophytes
If they feed on living things, then they're parasites

Mushrooms and toadstools, molds and mildews, yeast and rots
Are many kinds of fungi: some are good, some not
Lichen's really two things, living symbiotically
(Helping each other): algae and fungi

Mosses and liverworts, found in moist environments
Are simple, green, nonvascular, spore-producing plants

Mushroom development

VASCULAR PLANTS
(to the tune "Battle Hymn of the Republic")
Lyrics by Doug Eldon ©

Xylem carries minerals and water toward the sky
Phloem carries food on down and that's the reason why-
The most important cells are those producing both of them:
The cells of the cambium

Chorus: Vascular... Oh vascular plants
Vascular... Oh vascular plants
Vascular... Oh vascular plants
All have transporting tubes

Ferns and also horsetails are both vascular indeed
They reproduce by means of spores and rhizomes, not by seed
The spores come from the sorus underneath the fronds of ferns
Growing into a prothallium
Chorus

Gymnosperms have unprotected seeds which you may see
On the cones of conifers most often in a tree
Fir, pine, hemlock, cedar, spruce, redwood and juniper
Stay green throughout the year
Chorus

Gymnosperms with naked seeds and angiosperms with fruits
Absorb the water from the soil and held in place by roots
Food is made in leaves and then the stems help to transport
That food throughout the plant which they support
Chorus

Flowers may have sepals, petals, pistils, and stamen
Ripened ovaries are fruits with seeds inside of them
Seeds have coats and embryos that are indeed alive
With food to help them survive
Chorus

CO_2 and H_2O with light for energy
Glucose is made by chlorophyll and oxygen's set free
(Respiration is the opposite of this process)
It's called photosynthesis
Chorus

Plants need water, light and air with proper temperature
Space to grow, and minerals, they need them to mature
Tropisms are responses to stimuli, I'm told
By hormones they're controlled
Chorus

PROTOZOA
(to the tune of "Listen to the Mockingbird")
Lyrics by Doug Eldon ©

How they move is how you know
Into which group they will go
Here are four groups with some examples
So you understand how they are classified

Chorus: Protozoa, also called protista:
They're microscopic and are single-celled
Protozoa, also called protista:
They're microscopic and are single-celled

Amoebas are a sarcodine
Pseudopods to move are seen
False feet arrange the shape to change
And remember that the sarcodines are called:

Chorus

All around the paramecia
You can see some moving cilia
Tiny hairs are wiggling there
And remember that the ciliates are called:

Chorus

At the tail end of the Euglena
They will all have a flagella
The tail to whip is from the tip
And I know that the flagellates are called:

Chorus

Sporozoans live in a host
Movement's limited the most
Malaria's carried by a mosquito
The plasmodium is moved to a new host

Chorus

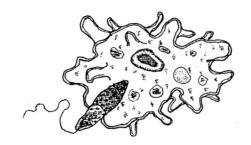

Amoeba

GENETICS

(to the tune of "Shortnin' Bread")
Lyrics by Doug Eldon ©

DNA is what they say
Made you what you are today
Like a program it will tell
What will happen in each cell

DNA is a molecule
Look real close, it's really cool
Like a ladder in a double helix
So simple yet so complex

Chorus: De-oxy-ribo-nucleic acid—
Talkin' 'bout genes and chromosomes
De-oxy-ribo-nucleic acid—
Talkin' 'bout genes and chromosomes

In chromosomes made of DNA
There is a particular way
Nitrogen bases can be seen
In special orders to make a gene

When organisms reproduce
Different genes are on the loose
Dominant and/or recessive
The offspring get traits the parents give

Chorus

Genetics is the study of heredity,
It may use probability
To predict just how likely
Results of genetic crosses will be

Chorus

Genetic engineers can use
Parts of different molecules
Changing DNA around
To solve a problem they have found

Chorus

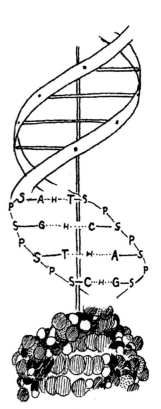

DNA molecule

VIRUSES
(to the tune of "Yankee Doodle")
Lyrics by Doug Eldon ©

Viruses cause many different infectious diseases
Influenza, common colds (with fevers, coughs and sneezes)
Rabies, viral pneumonia, and mononucleosis
Chicken pox, measles and mumps, AIDS, conjunctivitis

Chorus: Viruses are very small
Extremely microscopic
Few of them are good at all
From them the host may become sick

Viruses are not living, but are just tiny particles
Can't perform life functions; are not cells and don't contain cells
Reproduce only within a living cell and will take most
Of that cell's own processes to multiply inside a host

Chorus

In the core of viruses you'll find nucleic acids
Molecules which do control production of new viruses
RNA or DNA, under close inspection
Are surrounded by a coat of protein for protection

Chorus

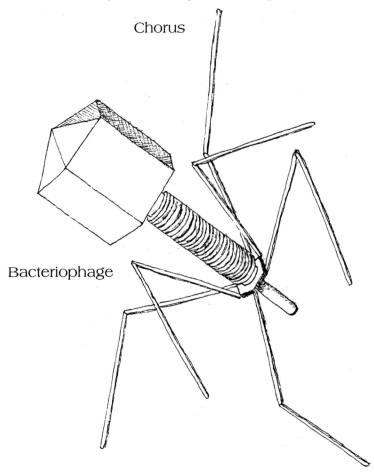

Bacteriophage

OH BACTERIA

(to the tune of "Oh Susanna")
Lyrics by Doug Eldon ©

Oh, lacking any nucleus, you do have a cell wall
You live in water, air and soil, and anywhere at all
You reproduce by fission, and you do so very fast
And under harsh conditions in an endospore you last

Chorus: Oh bacteria, though simple and so small—
Without you ecosystems would not function well at all

For decomposing things that die, a saprophyte we need
But some are parasitic, on a living host will feed
For taking nitrogen from air, and fixing it into
The soil for plants to use, I'm sure they're all grateful to you

Chorus: Oh bacteria, though simple and so small—
Without you ecosystems would not function well at all

In dairy products you have shown yourself a cultured friend
And to genetic engineers, your DNA you lend
You even help to fight diseases caused by your brethren
You make antibiotics which destroy or weaken them

Chorus: Oh bacteria, though only single-celled
A most important organism we have now beheld

Though most of you are helpful, in some of these mentioned ways
There are a few who have to do, a bit with some disease
Producing toxin or the cells attacking directly
Diphtheria, pneumonia, strep throat, tetanus, and TB

Chorus: Oh bacteria, though only single-celled
A most important organism we have now beheld

We do appreciate you and your praises we do sing
Yet some of you make life so hard with troubles that you bring
Our food you spoil, our crops you rot, our animals attack
With botulism, different rots, cholera and anthrax

Chorus: Oh bacteria, though only single-celled
A most important organism we have now beheld

Bacteria

CLASSIFICATION OF ANIMALS

These are the most commonly seen phyla, classes, and orders.

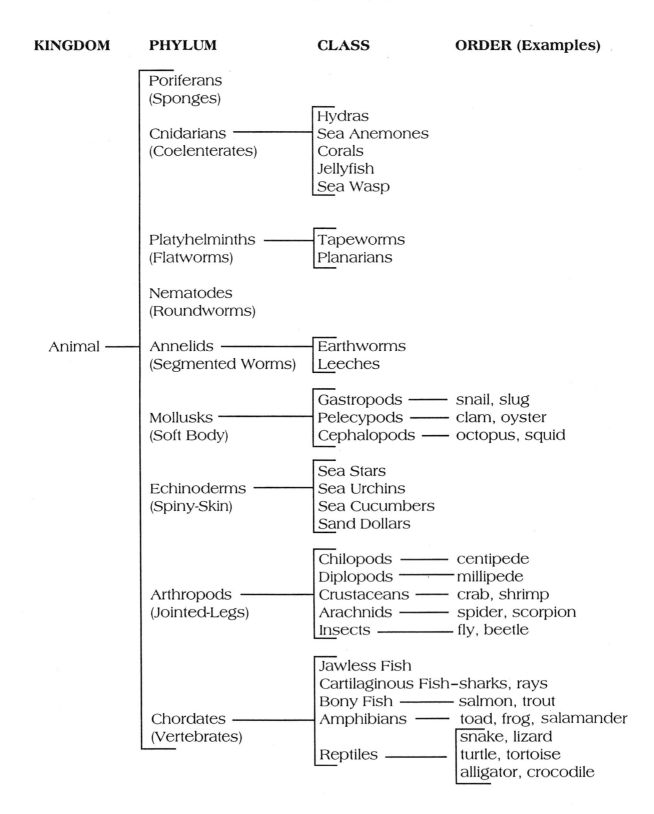

CAREER SEARCH

We hope you enjoyed learning and singing about life science! If you enjoyed a particular creature or subject, please use the listings below to find out what is necessary to prepare for a career in life science. These professional organizations have been contacted and are eager to share information concerning careers in their field of life science. They have pamphlets, brochures or booklets prepared to send you.

General Life Science
>Education and Human Resource Director
American Association for the Advancement of Science
1333 H. St. NW
Washington, D.C. 90560

>Dr. Pat Nellor Wickwire
American Association for Career Education
2900 Amby Place
Hermosa Beach, CA 90254-2216
Your request for career information will be forwarded to an association member in the particular science field requested.

Birds
>James Dean
American Ornithologist Union, Division of Birds
MRC - 116
Smithsonian Institution
Washington, DC 90560

Cell Biology
>American Society for Biochemistry and Molecular Biology
9650 Rockzell Pike
Bethesda, Maryland 20814

Vertebrates
>Public Relations
American Veterinary Medical Association
1931 N. Meacham Rd. Suite 100
Schaumburg, Illinois 60173 - 4360

Protozoa
>Lea Bleyman
Society of Protozoologists
Box A0506
Baruch College
17 Lexington
New York, New York 10010
Pamplet in progress.

Plants
>Iva Greenlee, Member Society Representative
American Phytopathological Society
3340 Pilot Knob Road
St. Paul, Minnesota 55121 - 2097

Insects
>Entomological Society of America
9301 Annapolis
Lanham, Maryland 20706 - 3115

NOTES

1. Matthews, Peter, editor. *The Guinness Book of World Records 1994*, p. 113.

2. *Ibid.* p. 110.

3. *Ibid.* p. 63.

4. *Ibid.* p. 88.

5. *Ibid.* p. 82.

6. *Ibid.* p. 133.

7. *Ibid.* p. 122.

8. *Ibid.* p. 60.

9. *Ibid.* p. 175.

10. *Ibid.* p. 176.

BIBLIOGRAPHY

Aylesworth, Thomas G. *The World of Microbes*. London: International Library, Collins Publishers, Franklin Watts, 1975.

Birch, Beverly. *Louis Pasteur*. Milwaukee: Gareth Stevens Children's Book, 1989.

Canby, Thomas Y. "Bacteria." National Geographic Volume 184, No. 2 August 1993. pp. 36 - 61.

Donlan, Edward F., Jr. *Adventure with a Microscope: A Story of Robert Koch*. New York: Dodd, Mead and Company, 1964.

Knight, David C. *Robert Koch Father of Bacteriology*. New York: Franklin Watts, 1961.

Mann, John. *Louis Pasteur Founder of Bacteriology*. New York: Charles Scribner's Sons, 1964.

Matthews, Peter, editor. *The Guinness Book of World Records 1994*. New York, Toronto: Bantam Books, 1994.

Metos, Thomas H. *Communicable Diseases*. New York: Franklin Watts, 1987.

Nourse, Alan E. *Viruses*. New York: Franklin Watts, 1976.

Prescott, Dr. Gerald W. *The Diatoms*. New York: Coward, McCann and Geoghegan, Inc., 1977.

Schwartz, George. *Life in a Drop of Water*. New York: The Natural History Press, 1970.

Shuttleworth, Floyd S., Herbert Zim. *Non-Flowering Plants*. New York: Golden Press, 1967.

Sterling, Dorothy. *The Story of Mosses, Ferns and Mushrooms*. New York: Doubleday and Company Inc., 1955.

Tames, Richard. *Louis Pasteur*. New York: Franklin Watts, 1990.

Weir, T. Elliott, C. Stocking, Michael G. Barbour. *Botany: An Introduction to Plant Biology*. New York: John Wiley and Sons Inc., 1970.

Wilcox, Frank H. *DNA The Thread of Life*. Minneapolis: Lerner Publications, 1988.

Wright, J., C. Coble, J. Hopkins, S. Johnson, D. Lahart. *Life Science*. New Jersey: Prentice Hall, 1988.

INDEX

A

aerie 32
AIDS 60
air bladders 28
air sacs 31
algae 34–37, 40–41
amoeba 49
amphibians 28
anaerobic 12
angiosperms 44, 47
annelids 21
anthrax 66
antiseptic 68
arachnids 23
arthropods 22
Aves 33

B

bacilli 64
bacteria 63–71
bacteriology 64
binary fission 49, 63
bioluminescence 37
biotechnology 55
birds 31–33
bivalves 21
botulism 70
brown algae 37
brown rot 38
bubonic plague 69

C

cambium 43
carnivores 23
cartilage 27
cells 13–14
cephalopods 21
chicken pox 59
Chilopoda 23
chlorophyll 14, 35, 37, 38
chloroplasts 14
cholera 67
chordates 27
chromosomes 14, 54
cilia 50
ciliates 50
cnidarians 19
cocci 64
coelenterates 19
coldblooded 27
communicable disease 57
competition 12
conifers 44
conjunctivitis 60
consumer 12, 64
crustaceans 22

D

deciduous 44
decomposers 64
decomposition 63
deoxyribonucleic acid, or DNA 54–55
diatoms 37
diphtheria 69
Diplopoda 23
dominant 53
double helix 54
dry rot 38
DPT 70
dysentery 70

E

echinoderms 20
ecosystem 64
endospore 63
epidemic 59
estivate 13
Euglena 50
evergreens 44
excretion 11
exoskeleton 22

F

feathers 31–33
fermentation 39
fertilization 45
fifth disease 61
fire algae 37
fire blight 71
fish 27, 28
flagella 50
flagellates 50
flowers 45
foot and mouth disease 61
frond 43
fruit 45
fruiting body 38
fungi 14, 35, 38, 39, 41

G

gastropods 21
genes 53
genetic engineering 55, 66
genetics 53–55
geotropism 46
germ theory 9, 68
German measles 59
germination 45
germs 57, 68
girdling 43
glucose 46
golden algae 37
green algae 37
gymnosperms 43

H

habitat 19
hepatitis 61
herbivores 23
heredity 53
hibernate 13
HIV 60
hormones 46
host 50, 65
hydrophobia 59
hypothesis 8

I

immunizations 68
incubating 32
infections 63
infectious diseases 57
infectious mononucleosis 59
inflammation 60
influenza 59
inherit 53
inoculation 68
insects 24
instincts 32
invertebrates 19–25
isopods 22

K

Koch, Robert 66

L

larva 24
Leeuwenhoek, Anton 63

leaves 45
lichen 40
Lister, Joseph 68
liverworts 40

M

malaria 50
mantle 21
measles 59
Mendel, Gregor 53
metabolic activities 11
metabolism 11
metamorphosis 24
microbes 7, 66
microbiology 64
migrate 32
mildews 39
MMR 60
molds 39
mollusks 21
Monerans 14
moss 40
mumps 59
Muscid domestica 16
mushrooms 39
mutualism 39
mycelium 38
mycorrhiza 39
Myriapoda 22

N

naiad 24
nematocysts 19
nematodes 20
nitrogen bases 54
nitrogen-fixing bacteria 65
nodules 65
nomenclature 15
nonvascular 34
nucleic acids 58
nucleus 14
nymph 24

O

organ system 16
organism 17
organs 16

P

pandemic 59
paramecia 50
parasite 38, 65
parasitic 20
parts of plants 45
Pasteur, Louis 7-9, 67

pasteurization 67
pelecypods 21
pertussis 70
petals 44
phloem 43
photosynthesis 45, 46
phototropism 46
plant growth 46
plants 34
plasmodium 50
platyhelminths 20
pneumonia 59, 70
poliomyelitis (or polio) 60
pollen 45
pollination 45
polliwogs 28
poriferans 19
powdery mildew 38
primary consumer 12
probability 54
producers 12, 64
prothallium 43
protist 14
protozoa 49
pseudopod 49
pupa 24

R

rabies 59
recessive 53
recombinant DNA 55
red algae 37
red tide 37
Redi 6
reproduce 11
reptiles 29
respiration 11, 46
rhizomes 43
root rot 38
roots 45
rots 39, 71
rubella 59

S

sap 34
saprophyte 38, 64
sarcodines 49
scavengers 23
science 5
scientific method 5-9
secondary consumer 12
seed 45
silica 37
smallpox 60
sooty mold 38
sori 43

Spallanzani 7
spirilla 64
spontaneous generation 6
spores 35, 43
sporozoans 50
stems 45
stimuli, stimulus 11, 46
strep throat 70
symbiotic 39
systematic 6

T

tadpoles 28
taxonomy 15
TB 67
terrestrial 22
territory 12
tetanus 69
tissues 16
tobacco mosaic virus (TMV) 61
toxin 70
traits 53
tropisms 46
tuberculosis 67
typhus 70

U

univalves 21

V

vaccines 68
variables 8
vascular 34, 43
vertebrates 27-29
viral diseases 59
viruses 58-61

W

wet rot 38

X

xylem 43

Y

yeasts 39
yellow fever 60